Aurora 7

The Mercury Space Flight of M. Scott Carpenter

Other Springer-Praxis books of related interest by Colin Burgess

NASA's Scientist-Astronauts
with David J. Shayler
2006
ISBN 978-0-387-21897-7

Animals in Space: From Research Rockets to the Space Shuttle
with Chris Dubbs
2007
ISBN 978-0-387-36053-9

The First Soviet Cosmonaut Team: Their Lives, Legacies and Historical Impact
with Rex Hall, M.B.E.
2009
ISBN 978-0-387-84823-5

Selecting the Mercury Seven: The Search for America's First Astronauts
2011
ISBN 978-1-4419-8404-3

Moon Bound: Choosing and Preparing NASA's Lunar Astronauts
2013
ISBN 978-1-4614-3854-0

Freedom 7: The Historic Flight of Alan B. Shepard, Jr.
2014
ISBN 978-3-319-01155-4

Liberty Bell 7: The Suborbital Mercury Flight of Virgil I. Grissom
2014
ISBN 978-3-319-04390-6

Friendship 7: The Epic Orbital Flight of John H. Glenn, Jr.
2015
ISBN 978-3-319-15653-8

Colin Burgess

Aurora 7

The Mercury Space Flight
of M. Scott Carpenter

 Springer

Colin Burgess
Bangor, NSW, Australia

SPRINGER-PRAXIS BOOKS IN SPACE EXPLORATION

Springer Praxis Books
ISBN 978-3-319-20438-3   ISBN 978-3-319-20439-0   (eBook)
DOI 10.1007/978-3-319-20439-0

Library of Congress Control Number: 2015942683

Springer Cham Heidelberg New York Dordrecht London
© Springer International Publishing Switzerland 2016

Front cover: *Aurora 7* launched with Scott Carpenter aboard, 24 May 1962. (Photos: NASA)
Back cover: Left: MA-7 astronaut Scott Carpenter prepares for his mission. (Photo: NASA). Right: A recent photo of Scott Carpenter at the Kennedy Space Center. (Photo: *collectSPACE.com*/Robert Pearlman)
Cover design: Jim Wilkie
Project copy editor: David M. Harland

Printed on acid-free paper

Springer International Publishing AG Switzerland is part of Springer Science+Business Media (www.springer.com)

Contents

There is one spectacle grander than the sea,
that is the sky; there is one spectacle grander
than the sky, that is the interior of the soul.

– Victor Hugo (1802–1885)

This planet is not terra firma. It is a delicate
flower and it must be cared for. It's lonely.
It's small. It's isolated, and there is no resupply.
And we are mistreating it. Clearly, the highest
loyalty we should have is not to our own country
or our own religion or our hometown or even
to ourselves. It should be to, number two,
the family of man, and number one, the planet
at large. This is our home, and this is all we've got.

– Scott Carpenter (1925–2013)

Foreword

In May 1959, a few weeks after NASA had announced the selection of the seven Mercury astronauts, a 23-year-old Air Force nurse began working at the Patrick Air Force Base hospital in Florida. Just six months later, before she'd really had a chance to settle in, the hospital commander, Colonel George Knauf, called her into his office and asked if she would like to consider working as a nurse for NASA's astronauts at nearby Cape Canaveral. Much to her astonishment, 2nd Lt. Dee O'Hara would find herself at the very heart of perhaps the greatest scientific endeavor ever undertaken, and doing one of the most enviable jobs in the world. She had become the trusted nurse for America's astronauts.

Although I wasn't really sure what the job entailed, I accepted. I really didn't know what to expect or what was ahead of me when I was sent to the Cape. I must also admit that I didn't really know what an astronaut was, and the only thing I knew about Mercury was that it was a planet, and also the stuff in a thermometer. It turned out that I was to set up the Aeromed Lab, an examination area for the astronauts in a building simply called Hangar S, and be with them as their nurse. I always felt the Mercury program was launched from there because we were all crammed into this one little hangar.

I was 23 years old back when I started work at the Cape, and admittedly not very sophisticated, so my first introduction to the astronauts came as a bit of an unexpected shock. Nobody had been assigned to a flight at that point, it was so early in the program, and this probably would have been January 1960. I happened to go down the hall one day and when I opened the door to the conference room they were all sitting there, with John Glenn seated on the table. I had no idea they were in the room and felt terribly embarrassed and even a little frightened that I had barged in on them. They were, after all, very famous people by then. I stammered an apology and backed out, slamming the door behind me, and fled back red-faced to my office. Then John Glenn, bless his heart, walked up and asked me to come back "and meet the guys," and said he would introduce me, which he did.

The premise behind me being there was that they wanted someone who would get to know the astronauts so well that she would work out whether they were really sick or not. The astronauts certainly were not going to tell the flight surgeon, Bill Douglas, because they knew the doctor had the right or capability of grounding them – the last thing these guys wanted. They were really fearful of doctors for this reason. So I made a deal with them: I said I would never betray them unless, in my opinion, what they told me would jeopardize them or the mission. In that case, I would have to report it to Dr. Douglas.

Back then, everything seemed to exist in Hangar S. It was essentially a long string of rooms off a narrow hallway which overlooked the floor of the hangar. We had a lab area, an exam room area, my little office, and then there was a large carpeted room where the space suits were kept. The suiting-up couch was in that room, where all of the suit check-outs were carried out. Then you went to the next room which was set up as a kind of a conference room for the astronauts. Past that was a little lounge area which was considered to be crew quarters, and in the next small room were bunk beds. If they were training late, or working in the capsule late, they could at least bunk there and sleep the night in privacy, and not have to drive the nineteen miles to a motel in Cocoa Beach, although they often used that distance to set hair-raising speed records in their powerful cars. Boys will be boys.

Initially, the astronauts were based at Langley Air Force Base in Virginia. During the week, they would fly down to the Cape in Florida for their testing, suit fittings, chamber tests, and so on. Then they would go back to Langley for the weekend. When it was launch time they would come down to Cape Canaveral and I would help them with pre-flight physicals, including height, weight, temperature, blood pressure and other tests.

Over the next few years, I served as the astronauts' nurse, tending to them and their families. I didn't really get to know their families well until I transferred to Houston, although I had met most of the wives when they'd traveled down to the Cape on rare occasions. Most had very small children at the time, so they stayed pretty close to home.

Once I had set up the Flight Medicine Clinic in Houston, the families would come to the clinic for medical care, in addition to their astronaut husbands. I became very close with the wives and knew them all intimately – they invited me into their homes. I am still in touch with many of them today.

They were wonderful days, but I always had my heart in my mouth whenever one of them was launched into space. You could feel the tension on everyone's part, because these men were entering the unknown and we didn't know what the heck was going to happen. It was both exciting and terrifying at the same time. Early on, I don't think any of us really thought about how historic these days would become. It was a job, and you just did your job.

I used to hear from people all over the world about how exciting they thought my job must be. I must admit I had the most ideal job in the world – I traveled, met movie stars, and got to associate with so many interesting people. I feel very fortunate to have been a part of a unique and exciting time in space history.

We all miss Scott Carpenter so much after he left us at the grand age of eighty-eight. The Scott I knew back in the Mercury days could see the beauty, the poetry or the emotion in so many things. He really was a poet, and a very sensitive, caring man.

Although it's only a small example of who he was, I won't forget the day he came into my office and said "Dee, the flint's gone in my lighter – do you have any flints?" I said I'd try to find one and after he'd gone I searched around and found a couple, which I taped

Scott Carpenter with astronauts' nurse Dee O'Hara. (Photo: NASA)

Scott Carpenter and Dee O'Hara raise a toast to Mercury astronaut Wally Schirra at his colleague's funeral in 2007. (Photo: Francis French)

under a piece of Scotch tape on his lamp. When I came in the next morning there was the piece of tape on my lamp, and underneath it was a note, and it said, "A kiss is in here for you." You see, that was Scott. It's hard to explain him to people, but that's a classic example of what he would say, and the gentleness and the sensitivity of the man. I've always had a very special spot for him.

I truly feel Scott did a wonderful job on his flight aboard *Aurora 7*. He got a lot of science done, a lot of the experiments that were thrust on him, and it's a shame he doesn't get enough credit for that. Yes, he missed his splashdown mark and scared everybody for a while. But the tragedy was when someone as powerful as Chris Kraft later came out and said, "I won't work with him again." That absolutely killed Scott's career. It was just heartbreaking; it shouldn't have happened. It was a very sad episode in an otherwise wonderful career, including his later exploits as a Sealab aquanaut. What an amazing person he was.

Over recent years, I caught up with Scott several times at different space events, and it was always a great pleasure to see him and to reminisce about old times and old friends. Scott especially loved talking to the children who came with their parents to the Astronaut Scholarship Foundation events. His eyes would light up as he asked them a number of questions about themselves and school. It was interesting to watch the children open up and talk to him without any shyness or hesitation.

We were all so sad when we heard the news of his passing, but like so many others I will always hold dear the precious memories of a man who gave so much to his nation, and yet retained the soul of a poet.

To paraphrase those unforgettable words to his great friend as John Glenn was about to be launched into orbit, *"Godspeed, Scott Carpenter."*

Dee O'Hara
Astronauts' nurse

Author's prologue

I first met Scott Carpenter in 1993 at the Association of Space Explorers' Congress in Vienna, Austria. After I'd asked him to sign my copy of his novel *The Steel Albatross* – which he did quite happily – I happened to mention that I had at home a photo of him riding a horse while stationed in Muchea, Western Australia, for the MA-6 orbital Mercury flight of John Glenn back in 1962. Immediately his face lit up and he smiled as he said, "Ah, yes – Butch." He remembered that horse well, and we fell into a conversation about the time he had spent in Muchea, and the people he'd met there. Some years later, Scott would recount by telephone the story of riding Butch in the Australian outback for a chapter in the book *Into That Silent Sea*, which I co-authored with Francis French.

Scott and I met again several times over the years at different venues, and it was always a true pleasure to sit and chat with a man who was a boyhood hero of mine. So much so that our older son was named Scott after him – a tribute he gladly repaid by sending me a surprise personal greeting for my son and his bride Melissa which I had the pleasure of reading out on the occasion of their wedding.

Although my enthusiasm for human space flight had kicked into gear during the much-delayed Mercury flight of John Glenn, the Mercury-Atlas 7 (MA-7) flight with Scott flying aboard *Aurora 7* was the first space mission I followed right from its inception, even when it was meant to be flown by another astronaut, and with the spacecraft bearing a different name. Some years later, I was able to view *Aurora 7* up close when it was on display at the Hong Kong Science Museum in Kowloon, never for one moment imaging that one day I'd not only get to meet Scott but befriend him and his daughter, Kris Stoever.

One truly unforgettable occasion came about on 26 January 2003 while I was visiting my long-time friend and co-author Francis French in California. We were delighted to have been invited to a private reception that afternoon at the Oakmont Country Club in Glendale by the Stoever family to mark the publication of Scott's autobiography *For Spacious Skies*, written with his daughter. After they had signed everyone's copies of their book, there was plenty of time to mingle with all the guests – many of them doyens of the space industry. Eventually it was time for Scott to make a short speech, thanking everyone

Scott Carpenter with the author at the Autographica show in Coventry, England, 2004. (Photo: Rex Hall)

The author enjoys a fun-filled conversation with Scott Carpenter at Spacefest V in Tucson, Arizona, May 2013. It would prove to be the last time they met. (Photo courtesy of *collectSPACE. com*/Robert Pearlman)

for attending. Midway through his oration he paused and casually mentioned that he would also like to thank another old friend who had just turned up. I turned around and it was a real pinch-me moment. The late attendee was Gordon Cooper. It would prove to be the first and only time that I had the opportunity to meet and talk with this other legendary Mercury astronaut, who sadly passed away the following year.

The last time I enjoyed a conversation with Scott Carpenter was at Spacefest V in Tucson, Arizona, at the end of May 2013 – in fact, on the 51st anniversary of his Mercury flight – and once again I enjoyed the chance to sit and chat with him as he kindly gave me a few words to use in an earlier book in this Springer series about his Mercury colleague Gus Grissom. Just five months later, I was deeply saddened to learn of Scott's death on 10 October, at the grand age of eighty-eight.

It was quite evident that Scott Carpenter (he told me he always disliked his given name of Malcolm) was cut from a very different cloth than that of his fellow Mercury astronauts. A dynamic pioneer of modern exploration, a superb athlete and test pilot, he was also a gentle and even poetic man; an experimentalist whose curiosity almost cost him his life on his only space flight and led to a well-documented falling out with certain influential members of the NASA hierarchy.

Nevertheless, his incredibly rich and diverse life certainly touched mine in so many ways, and I will be forever grateful to him and his family for allowing me to briefly touch on his own.

Acknowledgements

Bless their hearts, one and all, because a project such as this relies heavily on those who not only support the writing of the Mercury series of books, but are also willing and happy to contribute their memories and experiences.

While this particular book, out of necessity, relies heavily on numerous accounts previously written by, about, or recorded by Scott Carpenter – our amazing Dynamic Pioneer – it is the people who worked with him and knew him best that helped characterize this wonderful, adventurous man, who was blessed with the soul of a poet. I was fortunate enough to have met Scott many times over the years, and to have recorded a lengthy telephone interview with him back in 2002 – much of which was also mined for this book. He was certainly a much-loved man. Sadly, however, so many years have gone by since the unforgettable flight of *Aurora 7* in May 1962 that we have not only lost Scott Carpenter, but so many people deeply involved in his one and only flight into space. It made me doubly appreciative of those that I manage to locate who willingly dipped into the past for me and offered their accounts of that day in history, or revealed different aspects of the life and accomplishments of a greatly missed Mercury astronaut.

Many sincere thanks therefore go to Matthew Appelbaum (Mayor of Boulder, Colorado); Leigh Bartlow; Mike Blair; Ed Buckbee; Bill Cotter; Kate Doolan; Zachary Epps; Al Hallonquist; Joseph Hiura; Alan Humphries; Russ Kaufmann; Renee Mailhiot (Public Relations Coordinator) and Sarah Rosenbloom (Think Tank Volunteer) at the Chicago Museum of Science and Industry; Anne Mills (NASA Glenn Research Center); Bruce Moody; Michael Neufeld (Smithsonian National Air & Space Museum Space History Division); Robert Pearlman (*collectSPACE.com*); J. L. Pickering (*retrospaceimages.com*); Eddie Pugh; Alan, Alice and Tom Rochford; Scott Sacknoff (*Quest: The History of Spaceflight Quarterly*); and Patrick von Keyserling (Boulder Director of Communications).

Special thanks go to a dear and long-time friend who lived through those amazing years as the astronauts' nurse, Dee O'Hara, for agreeing to write such a wonderful Foreword to this book. Multiple thanks also to yet another friend and colleague of many years, Francis French, who ran his eagle eye over this manuscript – as he has done with so many previous works of mine – not only checking facts, making suggestions for improvement, and seizing upon typos for me, but also pointing out to an Aussie author where a certain word or phrase that I might have used would have had an American scratching his or her head in puzzlement.

My friendly "deputy" Tracy Kornfeld also read through the text and his input was greatly appreciated. The web site that he created with Scott Carpenter is chock-full of stories and biographical information by and about the man, and I encourage any reader to check out *www.scottcarpenter.com*.

Finally, I am truly and abundantly grateful to two extraordinary women who were not only close to the late Scott Carpenter, but offered to assist me in paying tribute to him through this book. I therefore humbly acknowledge the kind help and gracious encouragement of the Dynamic Pioneer's widow and daughter – Patty Carpenter and Kris Stoever.

And of course, I must thank the good folks who turned my manuscript into a book. Clive Horwood of Praxis Publishing in the United Kingdom; my superb copyeditor, fellow author and good mate David M. Harland; and Jim Wilkie, who produced the glorious cover art for this and all my Springer books. At Springer Books, New York, my effusive and ongoing thanks to the hard-working Maury Solomon, Senior Editor, Physics and Astronomy, and her incredibly helpful Assistant Editor Nora Rawn, who has worked miracles for me in so many ways.

1

A replacement astronaut

1960 was a monumentally busy and incredibly productive year for America's dynamic new civilian space agency, NASA (National Aeronautics and Space Administration), which had been formally established just two years earlier. But the agency faced an uncertain future. With an election looming later that year, incumbent president Dwight D. Eisenhower had shown little regard for the nation's space programs, taking the sword to NASA's budget for the following year. He also wanted to cut the second stage of the Saturn rocket program, consigning the space agency to a future based solely on flights in low Earth orbit, without the necessary support or funding to continue through to actually landing astronauts on the Moon.

Under the Eisenhower administration, NASA faced the bleak prospect of having the civilian space program basically put out to pasture after Project Mercury. Even his potential replacement in the White House, Republican candidate Richard M. Nixon, had very little enthusiasm for a costly civilian space program. Foreshadowing a drastic curtailment of human space activities, President Eisenhower voiced his opinion that, "further tests and experiments will be necessary to establish if there are any valid scientific reasons for extending manned space flight beyond the Mercury program."[1]

Despite this grim prognosis, midway through 1960 the Army Ballistic Missile Agency (ABMA) of the Redstone Arsenal, Huntsville, Alabama, became a part of NASA and was renamed the George C. Marshall Space Flight Center. In August the space agency successfully orbited *Echo 1*, an inflatable, aluminized balloon communications satellite, while on 19 December, in preparation for the agency's goal of a manned space flight, the first test flight of a Redstone rocket carrying a Mercury capsule was successfully completed.

In November 1960, America voted in a new president by the narrowest of margins. John Fitzgerald Kennedy became the youngest man ever elected to the office, and the first Catholic to occupy the White House. NASA still faced an uncertain future, knowing that during his campaign Kennedy had been disappointingly non-committal about the nation's non-military space program. Beyond making vague statements whenever the subject was raised, he was not at all convinced that sending humans to the Moon should form a major element of his forward strategies and policies. He had even stated that developing huge rockets was a colossal waste of taxpayer money.

© Springer International Publishing Switzerland 2016
C. Burgess, *Aurora 7*, Springer Praxis Books, DOI 10.1007/978-3-319-20439-0_1

President Eisenhower with (right) Dr. T. Keith Glennan, NASA's first Administrator, and Deputy Administrator Dr. Hugh L. Dryden. (Photo: NASA)

Before a session of Congress on 25 May 1961, President John F. Kennedy commits the United States to landing a man on the Moon before the end of the decade. (Photo: NASA)

History records that within half a year of taking office, and leading a nation stunned by the Soviet achievement of launching cosmonaut Yuri Gagarin into orbit, President Kennedy committed NASA and the United States to landing astronauts on the Moon's surface by the end of the decade. Instead of instigating a decline in the nation's space program, he used it to inspire the nation, and boldly set the country on what was arguably the greatest technological endeavor of all time.

FIRST ASTRONAUT CHOSEN

As 1960 rolled over into a whole new year filled with promise and excitement, albeit an uncertain future, NASA stood ready to decide who would become the nation's (and hopefully the world's) first space explorer – the man who would take America into space on a suborbital mission just a few months later.

The Mercury spacecraft for the historic suborbital shot was nearing completion at the McDonnell plant in St. Louis, Missouri, but NASA needed a pilot for the flight, and there were seven superbly qualified men to choose from. All were exemplary test pilots, and they had been undergoing intense space flight training since their selection back in April 1959. From the U.S. Air Force there were Capt. L. Gordon Cooper, Jr., Capt. Virgil I. Grissom and Capt. Donald ("Deke") Slayton; from the U.S. Navy, Lt. M. Scott Carpenter, Lt. Cmdr. Walter M. Schirra, Jr., and Lt. Cmdr. Alan Shepard, Jr. The seventh man, and sole Marine Corps representative, was Maj. John H. Glenn, Jr.

NASA's newly selected Mercury astronauts. From Left: Gus Grissom, Alan Shepard, Scott Carpenter, Wally Schirra, Deke Slayton, John Glenn and Gordon Cooper. (Photo: NASA)

The final decision on who would fly first fell to just one person, Dr. Robert Rowe Gilruth, at that time the appointed Director of NASA's Space Task Group. On 19 January 1961, just one day before the inauguration of the nation's newly elected president, Gilruth called the seven Mercury astronauts together into his office for an urgent meeting. Without revealing the exact nature of the exercise – although all seven quickly deduced its purpose – he handed each man a slip of paper and asked them to write down, in descending order, which person – obviously excluding themselves – they thought was best qualified to be given the job of making the first American space flight. Once they were done he collected the papers and instructed them to reconvene in the afternoon for an important announcement that would affect them all.

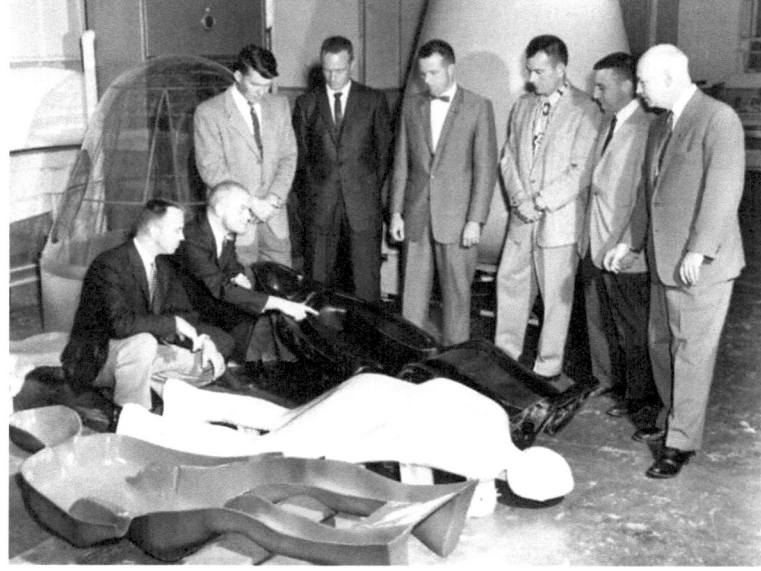

Dr. Robert Gilruth (far right) with the seven Mercury astronauts. (Photo: NASA)

Apart from poring over training reports and other indicators on the performance of the seven astronaut candidates, Gilruth was interested in assessing the men's peer ratings in order to determine whether their opinions coincided with his own. He knew that in the public's view the chief contender was undoubtedly the amiable red-headed Marine, John Glenn. And with very good reason – Glenn was a totally dedicated, rock-solid family man who spoke eloquently with pride and patriotism about the space program, and more than any of the other six candidates seemed to embody the ideals of a God-fearing, nation-loving astronaut.

Unbeknown to the public, however, Glenn didn't really stand a chance in a peer vote. In his eagerness to impress the NASA hierarchy and Mercury program officials with his abilities, and his famously pleasant countenance when dealing with the press, he had created a simmering resentment among his fellow candidates. Their major gripe came when he stepped over the line by angrily rebuking and lecturing them on their openly philandering ways with hordes of eager, willing women, which he insisted was not in keeping with the image they should portray to the public. The others resented being told by one of their own how to conduct their private lives, and this had a strong influence in their individual ratings. When it came to offering his own peer rating, Glenn unhesitatingly wrote Scott Carpenter's name at the top of his list. They had become good friends during the astronaut selection process. For his part, Carpenter was one of those who looked at a person's abilities and suitabilities, and this, combined with his friendship for the amiable Marine, meant that he placed Glenn at the top of his list.

This was no surprise, according to Ed Buckbee, NASA's public affairs officer at the Marshall Space Flight Center in Huntsville, Alabama from 1960 to 1970. "Scott was a kind, gentle and considerate man, never coming on as the cocky fighter pilot type. In my view, there always seemed to be a closeness between John Glenn and Scott. They jogged together when the other Mercury guys avoided running like the plague. They asked similar questions in briefings. They studied together and their families were close."[2]

Later that day the seven astronauts were called back into Gilruth's office, where he wasted little time in announcing that Alan Shepard would make the first suborbital flight. Next to fly he named as Grissom, followed by Glenn. Prior to their own missions, both men would serve as Shepard's backup pilots.

"I did not say anything for about twenty seconds or so," Shepard would recall. "I just looked at the floor. When I looked up, everyone in the room was staring at me. I was excited and happy, of course, but it was not a moment to crow." The other six, despite their disappointment, managed to put smiles on their faces as they congratulated him.[3]

Gilruth then told the seven men that he would not be making the announcement official for some weeks, in order to take the pressure off the primary astronaut during critical mission training.

Privately, Glenn was shocked and then furious at not being selected for the first American space flight. Or even the second. At a subsequent press conference in San Diego, he momentarily lost his renowned cool when asked who of the seven would be the first to fly. "We would like to get away from the 'first' aspect," he responded, a little testiness in his voice. "This is the beginning of many flights. Actually the second and third and fourth flights may accomplish far, far more scientifically than the first flight does. That first mission is going to be sort of an 'Oh, gee whiz, look at me; here I am, Maw!' type of deal, and you are probably going to get a limited amount of data back from it."[4]

Never one to sit back and take defeat lightly, Glenn even decided to question Gilruth's decision with the NASA administrators who were appointed following President Kennedy's installation in the White House. However they refused to overturn the decision and a sullen Glenn temporarily withdrew completely from the public view, later confessing that it took a while for him to bounce back and once again involve himself in the important tasks at hand.

"Those were pretty rough days for me," Glenn later admitted. "I guess I am a fairly dogged competitor, and getting left behind … was like always being a bridesmaid but never a bride."[5]

Although sharing Glenn's disappointment at the time, Scott Carpenter was far more philosophical about what he ultimately hoped to achieve in being an astronaut, as he explained to interviewer and space expert Ed Buckbee. "I think each of us had slightly different motivations. Some, I believe, wanted to be first. Some thought ahead to flying to the Moon and some thought of flying one day to Mars. My motivation was not much towards flying higher or faster or being first. I was motivated by curiosity to go to a place where man had never been and bring back understanding."[6]

Despite the competition between them, the seven men also recognized that they were involved in a dangerous research program and would rely heavily on each other, as Scott Carpenter explained, "Everybody wanted to take the first flight, whatever it was. We were all in a heated competition with each other. But the Three Musketeers came to my mind – we were all for one, and one for all. Truly we were. Although we struggled sometimes with petty difficulties, we were all one team where the program was concerned."

Scott Carpenter reporting for work outside the Mercury Control Center, Cape Canaveral. (Photo: NASA)

Carpenter outside the Mercury Control Center again, this time clad in his Mercury pressure suit. (Photo: NASA)

The peer review was an exercise often used as an evaluation tool in the military. When asked if he felt that Gilruth's use of the review had been a determining factor in the selection process for the first flights, Carpenter said it might have been.

"Bob made the decision [and] he probably took the ratings into consideration – he never told us what his decision-making process was. But I remember saying, in my peer review, that John was my first choice as backup, and conversely that if I had to be anyone's backup then I'd prefer being John's. Of course, in the end, John got the flight. I have an idea he named me as his backup of choice. So John got *two* things he wanted, and I only got one of mine!"[7]

GO FOR ORBIT

On 5 May 1961, Alan Shepard became the first American to fly into space, strapped into a compact spacecraft he had named *Freedom 7*, on suborbital mission MR-3 (a designation meaning the third launch of the Mercury-Redstone combination). Much to his and the country's chagrin, he had been beaten into space just three weeks earlier by Lieut. Yuri Gagarin of the Soviet Union, who had completed a single orbit of the Earth in a Vostok spacecraft.

During Shepard's flight, Scott Carpenter said that he and Wally Schirra were assigned the duty of chase plane pilots, observing the launch from high over the Cape in F-102 jets.

"For Alan's flight, Wally Schirra and I, in keeping with an old Air Force – Edwards [Air Force Base, California], as a matter of fact – practice of chasing every experimental flight with airplanes … Walt Williams from Edwards, highly placed in the administration in those days, thought we should chase Al Shepard's flight just because it was always done. So Wally Schirra and I were given some Air Force airplanes to chase Al's flight. We orbited and we couldn't stay close to the pad, because there were a lot of unknowns and dangers in those days that we didn't quite know how to cope with. But Wally and I were circling the pad, listening to the count, but at some distance, maybe 3 miles away. And Al took off, going straight up, and Wally and I never saw a thing! You can't chase a Redstone going straight up in a 102, so all we did is fly circles. And we came down and sort of said to each other, 'What happened?'"[8]

Gus Grissom followed Shepard two months later on the similar suborbital MR-4 mission. Both were successful ballistic flights, although Grissom's spacecraft was lost when the hatch on *Liberty Bell 7* blew prematurely and the capsule sank to the bottom of the Atlantic Ocean. John Glenn had acted as a backup pilot for both missions, and was next in line for his own suborbital shot. Then things changed.

6 August 1961 would prove a turning point for NASA, when a second Soviet cosmonaut was fired into space. This time the pilot was Gherman Titov, who successfully flew 17 orbits aboard the Vostok-2 spacecraft. Now that the United States was fully engaged in a full-on race to the Moon, announced that May by President Kennedy, Gilruth and his advisors saw little point in pressing ahead with further suborbital flights. The modified Atlas rocket was ready to launch astronauts into orbit, and for their part the astronauts were more than ready. It was time for another astronaut get-together in Gilruth's office.

The meeting between the seven Mercury astronauts and Robert Gilruth, now Director of the Manned Spacecraft Center, and his deputy, Walt Williams, took place on 4 October 1961, coincidentally the fourth anniversary of the launch by the Soviet Union of the world's first artificial satellite, *Sputnik*.

This time the news was far more welcome to a delighted John Glenn. He would fly the MA-6 (Mercury-Atlas 6) first orbital mission, then scheduled for December. Scott Carpenter was named as his backup pilot. The follow-on mission, MA-7, went to Deke Slayton, with Wally Schirra acting as his backup. Shepard and Grissom would have to wait for their second flights, while a deeply disappointed Gordon Cooper, the youngest member of the group, had yet to be given his first mission assignment.

Glenn's appointment to the MA-6 flight was not universally applauded within NASA. In 2013, Gus Grissom's brother Lowell put up for auction a letter Gus had sent to their mother in October 1961. In the letter he confided that he and his fellow Mercury astronauts resented Glenn being selected to become the first American to orbit the Earth. "The flight crew for the orbital mission has been picked and I'm not on it," he wrote. "Of course I've been feeling pretty low for the past few days. All of us are mad because Glenn was picked. But we expressed our views prior to the selection so there isn't much we can do about it but support the flight and the program." There may have been a bubbling resentment, but what the public saw in *Life* magazine and elsewhere in the media of the day was a unified team of astronauts eagerly preparing Glenn for his flight.[9]

Scott Carpenter poses alongside John Glenn. (Photo: NASA)

On 20 February 1962, following months of frustrating delays – mostly caused by poor weather conditions – John Glenn was launched into space aboard the *Friendship* 7 spacecraft on top of Atlas 109D and successfully completed a drama-filled three-orbit mission. During the flight he faced several problems with his automatic attitude control system, and then there were grave fears that his heat shield might have slipped, in which case the capsule would be exposed to the ferocious heat of re-entry and his life would end in the midst of a white-hot fireball. All around the world there was a massive sense of relief when it was announced that he had completed a safe and successful splashdown at the end of his mission.

HEALTH CONCERNS

In mid-March 1962, only weeks after the high-profile flight of John Glenn aboard his *Friendship 7* spacecraft, NASA was faced with something of a destabilizing crisis within the ranks of its seven astronauts when Deke Slayton received some truly devastating news; he was grounded, and had been removed from his Mercury space mission on medical grounds.

Deke Slayton with his custom-made pressure suit at the B. F. Goodrich plant. (Photo: NASA)

Donald Kent Slayton was born in Sparta, Wisconsin, on 1 March 1924, one of seven children to Charles Sherman Slayton and his second wife, Victoria Adelia (née Larson), both of Norwegian stock. He spent his boyhood on the family farm near the town of Leon, and graduated from Leon Elementary School at the head of his class. He first became interested in aviation matters while watching airplanes from Volk Field and Camp McCoy fly over their farm, wishing he was up there rather than pitching hay. He entered Sparta High School in a class of 180 students, initially taking agricultural courses but later changing to concentrate on physics, chemistry and mathematics in order to pursue a possible flying career. Along with his brother Howard and his sisters Beverly and Verna, young Don made the five-mile trip to school each day with Russ and Lloyd Harris in a Model T Ford the Harris boys had bought for $26.

In high school there was no doubting Don's fitness; he played on the Future Farmers of America basketball team for three years and ran for the school's track team all four

Undergoing checks of the Goodrich pressure suit. (Photo: NASA)

years. During track season, he often ran the five miles to school in order to improve his wind and legs.

On his 18th birthday, having graduated 16th in his class, Don enlisted as an aviation cadet with the U.S. Army Air Force and completed flight training in Texas, at the Vernon and Waco airfields. He thought for a time he might be disqualified from flying as he had lost the ring finger on his left hand in a farming accident when he was five (which would later concern him in his astronaut application), but the physicians cleared him and he entered the war as a pilot on B-25 Mitchell bombers with the 340th Bombardment Group.

He would eventually fly in both theaters of war over Europe and Japan, in all flying 63 combat missions – 56 over Germany and 7 over Japan.

Discharged at the age of 23 with the rank of captain, Slayton entered the University of Minnesota to pursue his college degree, but retained his membership in the Minnesota Air National Guard, flying T-6 trainers with the Air Force Reserve. Two years later, in 1949, he received his degree in aeronautical engineering. After working for the Boeing Company in Seattle, Washington, he was called back to active duty in early 1951, and decided to make the Air Force his career.

While stationed with the 12th Air Force Headquarters in Wiesbaden, Germany, he met and in 1955 married a civilian Air Force secretary from California named Marjorie Lunney.

Deke Slayton test-flying at Edwards Air Force Base, California. (Photo: U.S. Air Force)

From 1955, as an Air Force test pilot school student and experimental test pilot at Edwards Air Force Base, Capt. Slayton flew practically every available type of jet fighter in racking up 3,400 hours in the air, of which 2,000 hours were in jets. In April 1957 he and Marjorie celebrated the birth of their son, Kent Sherman Slayton. It was also during this time that he picked up the nickname by which he would be identified for the remainder of his life – Deke, formed by an elongation of his initials, D. K.

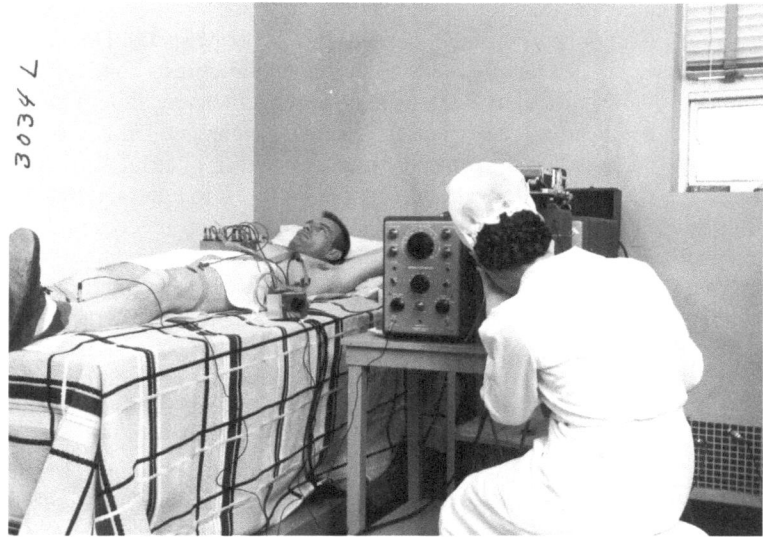

Slayton underwent a rigorous medical examination at the Lovelace Clinic in Albuquerque, New Mexico, prior to his selection by NASA. (Photo: Lovelace Respiratory Research Clinic)

In April 1959, after passing multiple examinations and a grueling set of mental and physical tests, Deke Slayton was named as one of NASA's seven Mercury astronauts. He and his family subsequently moved to Houston, Texas. But Slayton did have a problem he kept from everyone and that he would have to overcome as an astronaut – he could not swim. As Scott Carpenter related, "The one thing that was a standout about Deke was that he was a non-swimmer, and he didn't tell anybody that. He went through our training with the Navy SEALS, scuba diving and all that, and he never told anybody [about] that. He couldn't even swim. His wife used to talk about his practicing at home in the kitchen sink, inhaling through his mouth and exhaling through his nose. But that was a measure of his perseverance."[10]

Wally Schirra had an even more humorous take on his colleague's aquatic problem. "Deke Slayton was the best diver we had — he went right to the bottom! Gus Grissom and I had to pull him off the bottom, and help him tread water. In the movie *The Right Stuff*, they showed him cavorting with some girls in a water tank in a bar — the last person you'd really have put in that tank would be Deke Slayton! He was a farm boy out of Wisconsin. A river or a cistern was about the nearest thing he saw with water."[11]

The first inkling of a health problem for the 35-year-old astronaut came as early in his astronaut career as 27 August 1959. That day he undertook a centrifuge test, during which the g force was applied perpendicular to his chest, compressing the chest and making it difficult for him to breathe. The resultant electrocardiogram was something of a surprise to an attending Air Force physician on loan to NASA, Dr. (Lt. Col.) William Douglas, who had been assigned as the personal physician to the Mercury astronauts. When he reviewed Slayton's medical performance he noted the unmistakable signs of an irregular heartbeat – but occurring prior to the centrifuge test, not during the run. Slayton had earlier mentioned to Bill Douglas that he had "not felt up to par" that morning.

A recent viral infection might have caused the problem, so Dr. Douglas decided to hospitalize the astronaut in a nearby medical facility and have tests conducted. These tests revealed nothing of any particular concern apart from a minor variation in his blood count. His next step was a concerned phone call to Air Force colleague Dr. Lawrence ("Larry") Lamb. Douglas explained that Slayton had suffered an attack of atrial fibrillation – a slight irregularity of the heart that causes the top part of the heart to contract or twitch – which was possibly caused by a viral infection. He wanted to know if he could bring Slayton to see Lamb to obtain a more comprehensive cardiac evaluation and they agreed on 21 September. When Lamb reviewed Slayton's medical records from the Lovelace Clinic – the facility where exhaustive physical tests had been conducted in order to assist NASA in selecting the seven Mercury astronauts – he said, "I noted the same variation in the white blood cell count was present when he was examined before entering the astronaut program. It had been there all along and was of no real significance – so much for the recent virus infection theory."

Dr. Lamb determined there was nothing in Slayton's history to suggest a heart problem, but he was fully aware that atrial fibrillation could lead to a short episode of fainting, and the unknown factor in all of this was space travel – the dynamics and stress of launch and re-entry, and the largely unknown effects of weightlessness on the human physiognomy.

As Lamb later observed, "Orbiting an astronaut with atrial fibrillation could have made the United States the laughing stock of the world. If a catastrophe occurred, no one would know if it was really because of space flight conditions or from the heart irregularity. It would have been a very bad, ill-conceived experiment, costing millions of dollars. Congress was not all that enthusiastic about providing funds for the man-in-space program, either. Such a fiasco could have made NASA's effort to obtain funds that much more problematic."

The examination took place on the scheduled date, and as Lamb recounted, there was some particularly bad news in store for the anxious astronaut and his physician.

"It didn't matter how much I might have wished it to be otherwise; it was my painful duty to discuss the facts with Deke Slayton and Bill Douglas," Lamb recalled. "I always hated having to tell patients unwelcome news. At the completion of the examination, I explained to Deke that I doubted very much that he would be able to continue in the Mercury project. Clearly, he understood me. Bill Douglas was with him and most certainly, as a flight surgeon, he understood me. My opinion was a severe disappointment to Deke. Tears welled up in his eyes and ran down his cheeks, visible evidence that he did understand my remarks."

Despite his dire diagnosis, Larry Lamb explained that it was not his responsibility to make the final decision concerning Slayton's role in the nation's space program. Bill Douglas had asked him to express an honest medical opinion, which he had done. "All I needed to do was prepare a report and give it to him," Lamb stated. "[Douglas] would manage all matters thereafter. This was agreeable to me, since I was acting as a consultant to him in his capacity as the physician for the Mercury astronauts."[12]

Deke Slayton in discussion with executives from Convair/General Dynamics at San Diego, constructors of the Atlas rocket for Project Mercury. (Photo: San Diego Air & Space Museum)

Bill Douglas, while recording the test results in Slayton's record, did not feel as strongly as his physician colleague about disqualifying the astronaut from completing his space flight, and NASA's space program continued with two successful suborbital missions and Glenn's orbital flight giving massive impetus to the agency's Mercury schedule and future plans. Slayton was fully involved in mission training, and had even revealed that he had a name picked out for his spacecraft: *Delta 7*, the fourth letter of the Greek alphabet for the fourth Mercury flight, as well as what Slayton described as "a nice engineering term that described the change in number. Delta, as in delta-vee, [a] change in velocity."[13] Adding to his enjoyment at that time was an Air Force promotion to the rank of major.

Dr. Lamb, meanwhile, was growing increasingly concerned that no restrictive action had been taken against Slayton.

In his autobiography, published posthumously, Slayton wrote that Bill Douglas "and everybody else had agreed that it should have no effect on my status whatsoever. Everything was noted down in my records, and that was that." He consequently adjusted his diet, quit smoking and drinking, and increased his running schedule. "But it turned out that Lamb was also cardiologist to LBJ [President Lyndon Johnson], who had a whole history of heart problems .... Lamb raised hell at a pretty high level – which was how the rumors got started."[14] Thereafter, Slayton would refuse to ever speak Lamb's name.

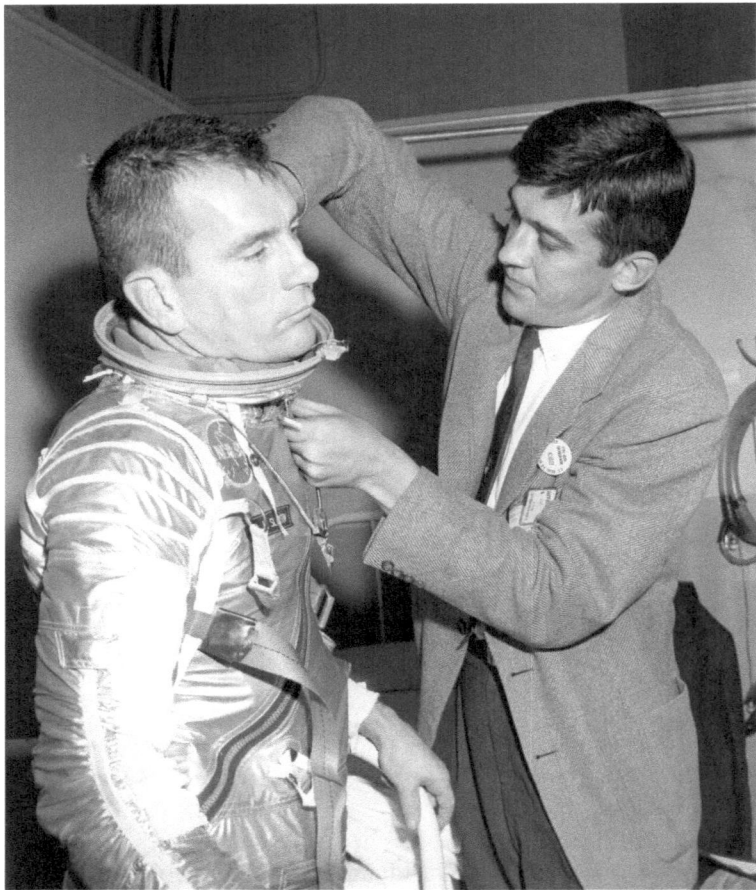

Deke Slayton being fitted into his custom-made pressure suit. (Photo: NASA)

Three weeks after John Glenn's Mercury mission, with his own flight just two months away, Slayton was completing some simulator work at the Cape along with his backup pilot Wally Schirra when he received an urgent phone call telling him to make his way to Washington. He didn't know the reason for the summons, but when he got there Bill Douglas met him and accompanied Slayton to the office of the surgeon general of the Air Force. "Bill made an opening statement about my history," Slayton recalled, "then they must have had every doctor in the place take a crack at me. It was at least twenty of them."

A training session with NASA suit technician Joe Schmitt. (Photo: NASA)

To Slayton's incredible relief, the consensus was that he was medically fit to fly. However the Secretary of the Air Force, Eugene Zuckett, suggested that the astronaut also undergo an examination by civilian doctors. The examination was by no means comprehensive; as Slayton later observed; they basically "poked and prodded" him and listened to his heart with their stethoscopes.

It was a surprisingly routine examination, and he had no reason for concern when it was finished and he had a meeting with NASA Deputy Director Hugh L. Dryden, "and just like that, Hugh Dryden told me I was off the flight. These guys didn't find any medical reasons to keep me from flying – what they said was as long as NASA has other pilots without this condition, why not fly them instead?

"I hadn't expected anything like this. I was just devastated."[15]

THE CHANGE IS MADE

At midday on Friday, 16 March 1962, looking the picture of abject misery, a distraught Deke Slayton was forced to endure a press conference at NASA Headquarters in Washington, D.C., along with NASA Public Affairs Officer, Lt. Col. John A. ("Shorty") Powers, Dr. Hugh L. Dryden, NASA Deputy Administrator, and Dr. Charles H. Roadman, Director of Aerospace Medicine at NASA's Office of Manned Space Flight. This followed the release the previous day of a space agency news bulletin in which it was officially announced that Slayton had been removed as pilot for the MA-7 mission. It was further announced that his place would be taken, not by his backup Wally Schirra, but by John Glenn's backup pilot, Scott Carpenter.

Col. Powers began the press by conference by stating, "As you all know, a decision was reached yesterday with regard to Major Don Slayton's role as pilot of the next manned orbital flight. Since that time there has been a great deal of interest from all members of the news media, and so we are attempting to respond to that interest and give you an opportunity to see that this fellow is not sickened to bed any place, but is live, hale and hearty, and to perhaps answer some of the questions that are on your mind."

Hugh Dryden was then asked to make a statement prior to any questions being asked. He said, "I might make one or two statements. First, the only decision that has been made is that Deke will not take the next mission in the MA-7. There are no decisions made to continue him for the future until there has been an opportunity for further examinations of this little defect that he has. All of us have similar things wrong with our bodies, I think. We want to understand more about the relationship and significance of this to future flights.

"Deke will continue in the program. He has a very important part to play in the next mission. We hope to get him back to work carrying on that part of the mission."

In response to a question concerning the circumstances under which the condition was initially discovered and why there was a delay until this time in making a decision, Dryden responded, "The defect was first discovered in 1959. At that time we had assurance from the medical people that this would not interfere with the mission. There has been no distinct change in this condition as far as I know. But in the continuing review by those who have the responsibility for the mission, it was decided wiser to make the change and to continue to take a little more time to assess the significance of this condition in relation to the stresses of orbital flight."

Charles Roadman was then asked if he could explain what the condition was. Before responding, he sought Deke's permission to do so, which was given. "With Deke's permission .... This is a diagnosis of paroxysmal atrial fibrillation. Paroxysmal, in simplest terms, is that it happens at indeterminate times. It is not continuously present, intermittent, are other words we might use to describe it.

"The atrial portion of the heart is one of the upper valve portions of the heart, the other portion being the pumping part, the ventricle. The atrium is more or less of a filling type receptacle for blood coming into the heart, going out. The fibrillation might be described in other terms as flutter or that type of increased movement. Another significant thing is the heart in its normal rhythm, so to speak, has an impulse mechanism which acts upon the atrial portion of the heart and also on the ventricle. When these do not operate, let's say, in normal sequence and input, you will develop an increased beat, if I may use that term, in the atrium. This is atrial fibrillation."

Col. Powers then added, "There are many unknowns still in this program; right now we feel that others who are sounder at the moment might be more appropriate. When we get to the point where there are not too many unknowns, this will not be a disqualifying condition. We expect to keep Deke flying right on down the road."

Eventually the questions were directed to Slayton. Asked when the fibrillations were first picked up and if there were any indications that it was related to the stress of training, he answered with both frankness and disappointment in his voice.

"As to the first part, the first time I was aware of it was in our first centrifuge program at Johnsville [Naval Air Development Center, Pennsylvania], which was also the first time I had had EKG leads on in quite some time, other than in physical exams. When the EKG leads were applied it was apparent I had this condition …. This was before I ever got on

NASA astronaut Lt. Cmdr. Malcolm Scott Carpenter, U.S. Navy. (Photo: NASA)

the wheel, before I ever got on the vehicle. It was not as a result of being on the wheel. Of course, I have been conscious of it at various times since that point. Up to that point I was not aware of it … it does not affect my performance in any way. I can do everything with it that I can without it, in terms of stress exercise and so forth …. In my opinion it isn't anything unhealthy; it's just like having one blue and one brown eye."

At this, Dryden felt a need to interject in support of the astronaut. "Let's make it clear. Deke is ready to go, as far as I am concerned."

Later in the lengthy press conference, Slayton was asked when and where he was told of the decision, and how he felt about it.

"As Dr. Dryden said, when this thing came up for re-evaluation two or three days ago, I was told I was being re-evaluated. I came up here yesterday to meet the latest board of cardiologists and I was told immediately after they had made their decision. Of course my feelings, I think, are very obvious. I am damned disappointed. Let's face it. You feel the same way you'd feel about anything when you get shot out of the saddle unexpectedly."[16]

TAKING UP THE CHALLENGE

Acting on the advice of Walt Williams, then head of NASA's Space Task Group, Robert Gilruth had decided to name Carpenter as the MA-7 pilot, with Schirra assigned as his backup. As Williams explained to Slayton's biographer, Michael Cassutt, "Everybody knew there was an outside chance that something would crop up to keep Deke from flying that mission – that medical condition, which we'd known about for a couple of years by that point. To be fair, there were a lot of operational questions you had to deal with. For example, what do you do if he starts fibrillating on the pad? Do you scrub the launch or go ahead? But we were prepared to deal with them because we thought Deke was the best choice to follow John Glenn in Mercury.

"But it was awful the way it happened. I don't remember Deke being angry as much as he was hurt – partly for the way it was handled."

When Hugh Dryden had asked Williams which astronaut should step in and take over as mission pilot, it came down to a choice between Schirra and Carpenter.

"I figured that MA-7 was likely to be more a repeat of John's flight than anything groundbreaking, so why not give it to Scott, since he had already trained for something pretty similar? We were thinking about a seven-orbit flight later in the year, and that would be perfect for Wally. So I was the one who made the decision to replace Deke with Scott."[17]

Initially, it was news that did not please Schirra at all. He argued with some justification that as Slayton's alternate *he* should have automatically slipped into the role of prime pilot. Gilruth patiently explained that Carpenter had served as Glenn's backup for more than five months and had logged more training hours for an orbital mission than either Slayton or Schirra. As it turned out, the decision to ground Slayton gave Schirra an even better assignment to the MA-8 mission later that year, with Gordon Cooper as his backup. Schirra would later write in his memoir, *Schirra's Space*, that he did his best and worked hard on the MA-7 mission, but "I don't think anyone knew how angry I was." He would add about

Deke Slayton, "Those were not happy days. If I was at a low ebb, Deke was touching bottom. And our doctors, Bill Douglas and another surgeon named Bill Augerson, were bitter. They had tried to seek second, third and fourth opinions on Deke's heart condition from other doctors around the country, and all they got were wishy-washy answers. That's the trouble with a lot of doctors. They're afraid to commit themselves."[18]

Clad in their silver pressure suits, Wally Schirra and MA-7 prime pilot Scott Carpenter practice water egress techniques. (Photo: NASA)

Flight Director Chris Kraft was incensed over the decision. As he later wrote in his memoir, *Flight*, "If they'd let him fly, he would have been the next American hero. Instead he was dumped. Slayton was the victim of overly fretful doctors and a NASA hierarchy that turned timid when it should have been bold. What happened to Deke Slayton shouldn't have happened to a dog."[19]

Meanwhile, the news of his late assignment to MA-7 had also caught Carpenter by surprise. He was aware that only a third of the preparations for the MA-7 flight had been completed by that time. As he recalled in the astronaut book *We Seven*, Deke Slayton had been working on detailed procedures with his *Delta 7* spacecraft when the news suddenly hit.

"Deke was so wrapped up in his work and so eager to go that one of the saddest things I have ever had to do in my life was to take over his assignment … when a board of doctors

Astronaut physician Dr. Charles ("Chuck") Berry sits alongside Deke Slayton at a later press conference. (Photo: UPI Telephoto)

finally ruled that he could not make the MA-7 flight. We had all known about the flutter in Deke's heart, but we all believed – and Deke had proved it to our own satisfaction – that it was a minor aberration which would have no effect on his role as a pilot. Deke had always performed as well as any of us during our strenuous training, and we were convinced this was an insignificant detail. But once more the top management was being cautious and Deke suffered a deep disappointment. I was happy to plunge back into work and prepare for my own flight, but I was sorry that my assignment was tinged by the great disappointment that everyone, including me, felt for Deke."[20]

Once Carpenter had been assigned the MA-7 flight – by "default" as he called it – he knew he had just eleven weeks to train for the flight. Among a host of responsibilities that had fallen his way was what substitute name he would now bestow upon his spacecraft.

He finally settled on *Aurora 7*. As he later explained, "I named it *Aurora 7* because I saw it as a celestial event, and the *aurora borealis* is a celestial event. I liked the sound of it and the celestial significance.

"First of all, let's go into '7'. Al Shepard started that with *Freedom 7*, and the press caught that and said, 'Isn't that nice of Al to name his capsule something 7 in honor of the seven astronauts as buddies.' And everybody believed that. The fact of the matter is that he named it 7 because it was Capsule Number 7 off the line [at McDonnell]. People didn't know that, but since everybody wanted to match Al's largesse, Gus had *Liberty Bell 7* and John had *Friendship 7*. So I had to do something with 7 in it, with *Aurora 7*.

"But people back home in Boulder [said], 'Wasn't that nice of Scott to name his capsule *Aurora 7* for the fact that he was born and raised in a house in Boulder on the corner of Aurora and Seventh Streets.' So I give you the real reason behind Aurora, but the people from Boulder don't believe it."[21]

Artist Cece Bibby talks to Scott Carpenter as she paints the Aurora 7 logo onto the side of his spacecraft. (Photo: NASA)

Carpenter's daughter and biographer Kris Stoever added to her father's comments. "The childhood connection, for my father, with *Aurora* and *Seven* is indelible and indisputable. Even more indelible for my father was the sight of celestial or other mysterious visual phenomena – he loves the numinous. It fires his considerable imagination. As it happens, this love, in addition to his 20–10 eyesight and scientific temperament, suited him ideally for observation-rich duties required in particular of the flight he was given. He appreciates such things, and ponders them, more than I can possibly convey in words.

"Also, recall that his flights near the North Pole, during the Korean War, brought him time and time again in view of the *aurora borealis*. He wrote home about them to my mother. Remember, too, that many of the early flights and rockets and programs were taking their names from classical or mythological themes or persons. Aurora was the goddess of the dawn, and Dad said at the time that he felt we were at the dawn of a new age, space exploration, and *Aurora*, the name, was one way to declare that belief. Interesting, too, that among the manly names being chosen and applied – Atlas, Mercury, Thor – my father wasn't afraid of honoring a gentle, light-bringing female deity."[22]

Like John Glenn before him, Carpenter recruited the talented and delightfully feisty Chrysler Aerospace artist Cece Bibby to design and hand-paint the *Aurora 7* insignia on the shingled exterior of his capsule. Chrysler was a subcontractor for NASA, and Bibby's graphic arts department was located just across the road from the astronauts' office. She had been responsible for working on everything from instruction manuals to providing artwork for NASA publications before completing the *Friendship 7* logo on Glenn's capsule. For *Aurora 7*, Bibby used a cluster of multi-colored rings to depict the *aurora borealis*, which had jagged edges to imitate the movement and electric feeling of the lights. Aesthetics aside, there was also an experiment involved in painting the logo, with Bibby using different brands of paint to determine which one would better survive the intense heat of re-entry.

The *Aurora 7* logo. (Photo courtesy of *scottcarpenter.com*)

Despite his disappointment, Slayton pitched in to help organize the MA-7 flight, now under the command of Scott Carpenter. But he believed Carpenter made a bad mistake by trying to take on too much on what was to be just a three-orbit mission.

"I had worked to keep the experiments to a minimum," Slayton said in his 1994 memoir, "since there was a hell of a lot we still hadn't demonstrated with Mercury, such as a reliable flight control system.

"Scott had a different perspective. He was always at home with the doctors and scientists – I think he was genuinely curious about the things that interested them. But it bit him in the ass during his flight."[23]

As Wally Schirra later noted, "There was one positive result to Deke's grounding. The other six of us decided we'd ask him to be our leader, the chief astronaut. There had been rumors about NASA bringing someone in to oversee the astronaut office, and we feared it would be a retired admiral or general, possibly even Warren North, the 'in-house' NASA astronaut, who definitely wasn't one of us. What we wanted the least was somebody who would outrank us and issue orders in a military way. We wanted someone who knew us, who trained with us. Deke was the one and only choice. Sure, we were acting out of sympathy. Deke had been through hell. But we were proposing him as our leader out of respect not pity."[24]

Deke Slayton in the Cape blockhouse for Gus Grissom's MR-4 mission. (Photo: NASA)

Gordon Cooper agreed with Schirra. "Deke was ready to quit, and the rest of us were so upset that we came close to resigning en masse, which would have left America's space program with no astronauts. We finally talked Deke into staying on as head of the astronaut office, a new post in which he would represent our mutual interests. It wasn't a charity appointment. At the time, there was serious talk of bringing in a general or admiral to fill the new post, but we all agreed that we didn't need some outside weenie coming in and telling us what to do."[25]

Robert Gilruth, Director of Houston's Manned Spacecraft Center, gave his blessing to the proposal, and Deke Slayton was named chief astronaut. He would prove many times over to have been the right person for the job. Despite his ongoing dismay at not being permitted to fly into space, he nevertheless carried out his duties with great and memorable distinction.

Slayton never gave up on his dream of flying in space. Accompanied by Dr. Charles Berry he traveled to Boston to see the famed cardiologist, Dr. Paul Dudley White, but sadly White agreed with the other cardiologists. A year later, he had to endure yet another major blow when the Air Force decided he could no longer remain on flying status with the service and would be grounded "when a report of medical examination is received as required by Air Force regulations." Faced with this miserable prospect Slayton submitted a letter of resignation by mail to the Air Force on 11 October 1963, with a request that it take effect on 20 November before the service could ground him. He said at the time that he felt he would be able to play a more valuable role in the space program as a civilian, because he dealt with both civilian and military astronauts in his role as chief of the astronaut office.

In April 1967, Dr. Berry placed Slayton on the pharmaceutical agent quinidine, which prevents fibrillation attacks. But two years later, in August 1969, he voluntarily stopped taking any drugs due to a hectic schedule that meant he was not taking them at the prescribed times. Although the fibrillation returned, he decided to stay off the cardiac medication. He had his last fibrillation episode in July 1970 and to his knowledge never had another. In August the following year Charles Berry arranged for a further examination, this time by Dr. Harold Mankin, a cardiologist at the Mayo Clinic. There were two key tests carried out simultaneously, and the more important of the two – a coronary angiogram – had come into its own in the decade since Slayton had been grounded. The test proved that he did not have a coronary artery disease. Now medically cleared, he was placed back on active astronaut status in January 1973.

Deke Slayton prior to his first and only space mission; the ASTP "handshake in space" flight in 1975. (Photo: NASA)

Finally, in July 1975, Deke Slayton achieved his greatest ambition at the age of 51, when he operated as a crewmember on the Apollo-Soyuz Test Project (ASTP) flight, spending nine glorious days in Earth orbit.

Aboard – just in case – was some quinidine.[26]

REFERENCES

1. Charles Murray and Catherine Bly Cox, *Apollo: The Race to the Moon*, Simon & Schuster, New York, NY, 1989
2. Ed Buckbee email correspondence with Colin Burgess, 26 December 2014
3. Colin Burgess, *Freedom 7: The Historic Flight of Alan B. Shepard*, Springer-Praxis Publication, Chichester, UK, 2014
4. Francis French and Colin Burgess, *Into That Silent Sea: Trailblazers of the Space Era, 1961–1965*, University of Nebraska Press, Lincoln, NE, 2007
5. *Ibid*
6. Ed Buckbee with Wally Schirra, *The Real Space Cowboys*, Apogee Books, Ontario, Canada, 2005, pg. 14
7. Scott Carpenter telephone interview with Colin Burgess, 18 December 2002
8. Scott Carpenter interview with Michelle Kelly, JSC Oral History program, Houston, Texas, 30 March 1998
9. *The Guardian*, website article, "Astronaut Gus Grissom in 1961: 'All of us were mad John Glenn was picked'", online at: *http://www.theguardian.com/science/2013/nov/21/gus-grissom-apollo-1-astronaut-john-glenn-letter-nasa*
10. Scott Carpenter interview with Michelle Kelly, JSC Oral History program, Houston, Texas, 30 March 1998
11. Francis French interview/article with Wally Schirra, "I worked with NASA, not for NASA," 22 February 2002. Available at: *http://www.collectspace.com/news/news-022202a.html*
12. Lawrence Lamb, *Inside the Space Race: A Space Surgeon's Diary*, Synergy Books, Saint George, Utah, 2006
13. Donald Slayton and Michael Cassutt, *Deke! U.S. Manned Space from Mercury to the Space Shuttle*, Forge Books, New York, NY, 1994
14. *Ibid*
15. *Ibid*
16. NASA News Release 62–67, *Pilot Change in Mercury-Atlas No. 7*, NASA Headquarters, Washington, D.C., 16 March 1962
17. Donald Slayton and Michael Cassutt, *Deke! U.S. Manned Space from Mercury to the Space Shuttle*, Forge Books, New York, NY, 1994
18. Walter M. Schirra, Jr. with Richard N. Billings, *Schirra's Space*, Quinlan Press, Boston, MA, 1988
19. Chris Kraft, *Flight: My Life in Mission Control*, Penguin Putnam, Inc., New York, NY, 2001, pg. 163
20. M. Scott Carpenter, L. Gordon Cooper, Jr., John H. Glenn, Jr., Virgil I. Grissom, Walter M. Schirra, Jr., Alan B. Shepard, Jr., Donald K. Slayton, *We Seven*, Simon and Schuster, Inc., New York, NY, 1962
21. C-SPAN Oral History, uploaded to YouTube 8 March 2011. Online at: *https://www.youtube.com/watch?v=iBslf6SV2RU*
22. Kris Stoever quote taken from Francis French and Colin Burgess, *Into That Silent Sea: Trailblazers of the Space Era, 1961–1965*, University of Nebraska Press, Lincoln, NE, 2007
23. Donald Slayton and Michael Cassutt, *Deke! U.S. Manned Space from Mercury to the Space Shuttle*, Forge Books, New York, NY, 1994
24. Walter M. Schirra, Jr. with Richard N. Billings, *Schirra's Space*, Quinlan Press, Boston, MA, 1988
25. Gordon Cooper with Bruce Henderson, *Leap of Faith: An Astronaut's Journey Into the Unknown*, HarperCollins, New York, NY, 2000
26. Mark Bloom, *The Citizen* newspaper, Ottawa Canada, "Heart almost ended career," issue 10 July 1975, pg. 72

2

From Colorado to the Cape

Malcolm Scott Carpenter (who always said he disliked his first name) was born in Boulder, Colorado, on 1 May 1925. He would be the only child of Florence Kelso (née Noxon), who went by the preferred name of Toye, and Dr. Marion Scott Carpenter, a chemist.

The future astronaut was 13 generations removed from his pioneering ancestor, William Carpenter, an emigrant whose family had lived in Shalbourne Parish, Wiltshire. In May 1638 William sailed from Southampton, England aboard the *Bevis of Hampton*, along with his wife, their four children, and his father, also named William. Ironically enough, the father and son were both carpenters by trade. From the ship's records it seems the elder Carpenter died on the journey across to New England. The family eventually settled in Rehoboth, Bristol County, one of the oldest towns in Massachusetts.[1]

For the first two years of Scott Carpenter's life the family lived in New York City, where his father had been awarded a post-doctoral research post at Columbia University. But in the summer of 1927 he returned to Boulder with his mother, who had contracted tuberculosis and was advised to leave New York for a healthier climate if she was to recover.

As he grew up, young Scott seldom saw his father, who was working and studying as a research chemist in New York and had little time for an ailing wife and an infant son living more than halfway across the country.

At a time when she should have been enjoying her own youth and being a mother to her little son, Toye Carpenter eventually became so ill that she was bedridden at the home of her parents, Victor and Clara Noxon, who lived in a brick house on the corner of Aurora and Seventh Streets, just outside Boulder. It had a huge back yard, which Scott loved. Over the next eleven years, as they cared for their daughter, Scott was mostly raised by his maternal grandparents. Even though they had seven grown children of their own, Vic Noxon, then editor of the *Boulder County Miner and Farmer* newspaper, developed a close relationship with his young grandson, who had acquired the nickname Buddy.

© Springer International Publishing Switzerland 2016
C. Burgess, *Aurora 7*, Springer Praxis Books, DOI 10.1007/978-3-319-20439-0_2

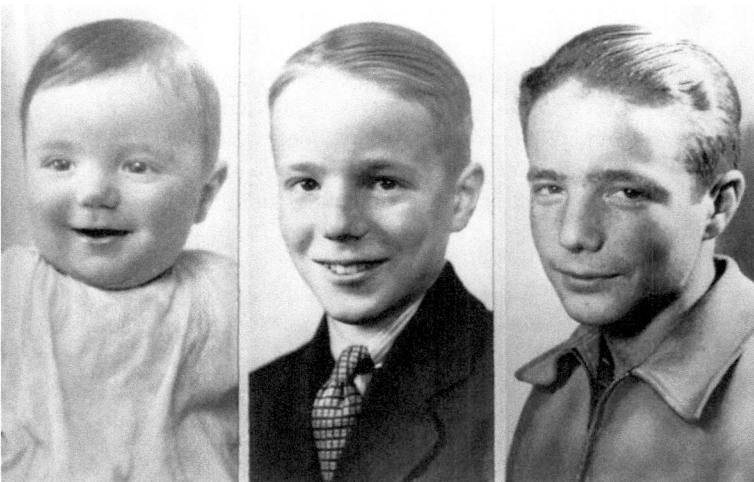

The first few years of Scott Carpenter's life are reflected in these photos. (Photos: M. Scott Carpenter Collection, Archives, University of Colorado at Boulder Libraries)

A recent photo of the much-modified Noxon home on the corner of Aurora and Seventh Streets, where Scott Carpenter grew to adulthood. (Photo courtesy of Zachary Epps, Boulder Real estate)

THE GREAT OUTDOORS

Because he had grown up exploring the great outdoors around the foothills of the Rocky Mountains in Clear Creek County, Vic Noxon would often regale Scott with stories about the days of gold miners and Native Americans. Scott was enthralled by these tales – his first contact with adventure. Noxon would take the boy hiking through the woods, teaching him about the wonders and dangers of the wilds, training him in survival skills, the art of setting traps, the importance of a well-honed knife, and how to fend for himself. He also gave Scott his first job, delivering newspapers.

"He was my role model, my father figure and the most influential adult male in my entire life," Carpenter told interviewer Woody Paige. "He was a great man. I had no father at that time; he was gone. My parents were split, so I didn't have a father at home. But I lived with my grandfather, and I think of him as the best father figure anybody could ever want. I was lucky. I wasn't deprived of a father; I had a marvelous father figure at home in my maternal grandfather."[2]

Meanwhile Scott's mother had become so ill with tuberculosis that she needed constant care, and in early 1931 she was signed into a nearby health sanatorium. Eight years later, in 1939, Scott's grandfather died of a stroke while sitting at his editorial desk in the newspaper office. Although he was still living with his grandmother Clara, 14-year-old Scott became introspective and something of a loner, preferring his own company and developing a sense of self-reliance. His mother once recalled that he took care of things himself without the help of others, and loved nothing more than riding his horses Lady Luck and Toby on Table Mesa, high above Boulder.

Scott and his beloved Lady Luck at the Noxon's Boulder home. (Photo: M. Scott Carpenter Collection, Archives, University of Colorado at Boulder Libraries)

Eventually Scott's mother recovered enough that she was able to return from the sanatorium, but she remained frail. Meanwhile, he would only see his father for a short time during summer breaks from school. Years later, in 1945, following a lengthy separation, his parents would end their marriage in divorce. There was further sadness in Scott's life when his grandmother Clara passed away 18 months after her husband, and Scott often had to fend for himself.

A popular student at high school, he was also a standout athlete, lettering in gymnastics and serving as president of the ski club. He sang in the glee club, acted in school plays, and served as acolyte at St. John's Episcopal Church where he had been christened.

In Carpenter's own words, he was something of a hell-raiser in his youth, and was rapidly on the way to becoming a loser in life. "I stole things from stores and I was just drifting through, sort of no-good," he told *Life* magazine's Loudon Wainwright.[3] He was generally living a restless existence and even quit his high school football team because he realized he had little interest in learning the plays. Armed with his driver's license from the age of 16, he was often involved in risk-filled endeavors, as observed in the article Wainwright wrote on Carpenter's often erratic upbringing:

"His independence from adult authority gave him stature, and this stature increased when he got his car. The car was a 1934 Ford coupé, which he paid for by selling a clarinet he'd discarded years earlier. He installed mica-pack mufflers to make the twin tail pipes roar, re-hinged the trunk, got some old seats out of a junkyard for a rumble seat, filched a pair of tiger-eye taillights, sawed the gear shift lever off so that the knob was only a few inches from the floor, and had himself a marvelous and dangerous toy."[4]

A BRUSH WITH DEATH

Graduating from high school in 1943, Carpenter made plans to enter the Navy's V-12a flight training program, having decided to become a naval aviator after seeing a film that would change his life.

"*Wake Island*, the movie, made me decide to be a naval aviator. Not an aviator. A naval aviator," he said. "I had a scrapbook that I kept as an 8-year-old that has airplanes in it. I was fascinated by airplanes. In that scrapbook, I also had pictures of the sleeve insignia of naval officers. The movie crystallized it for me for two reasons – it involved flying, and it also involved defending the country."[5]

Before turning eighteen, Carpenter traveled to Twelfth Naval District Headquarters in San Francisco hoping to qualify for the program, and was successful; he was enrolled as a U.S. naval aviation cadet. With no real need for his souped-up car while in the Navy, Carpenter arranged for his mother to take care of it. "He left the car with me," Toye once said in an interview. "I was the envy of every teenager in town."[6]

In September that year, now aged 18, Carpenter began basic training in the V-12a program at Colorado College, and in two years he had advanced to pre-flight in California.

The next move on the ladder was to primary flight training in Ottumwa, Iowa, where he was hoping to qualify for combat duty with the Navy. But the war came to an end before he won his wings. "The war was over before I learned to fly," Carpenter later recalled. "I only had eight hours in the Stearman N2S. So I returned to Boulder, re-entered school at the University of Colorado, and nearly got myself killed in a car accident."[7]

Scott Carpenter and his good friend Bill Todd working on "Bessie," Scott's lovingly rebuilt 1934 Ford roadster. (Photo: Carpenter family)

Just after one o'clock on the morning of 14 September 1946, after giving a friend a lift home, Carpenter was driving back to Boulder in the roadster he had nicknamed "Bessie." It had been a long and exhausting day, and he was fighting to stay awake, playing loud music, singing and smoking. Before he knew what was happening he had drifted off to sleep while traveling at what was later estimated to have been around 80 miles an hour. The Ford coupé swerved off the road, smashed through some guard posts on a bend and plunged 100 feet off the mountain road into a culvert. He was thrown clear, rolled to a stop on the ground, and – badly injured – began slipping in and out of consciousness. Eventually, covered in blood, he managed to crawl up to the road, where he passed out once again. Fortunately a couple happened to notice him on the roadside as they drove past, turned around and rushed him to Boulder Community Hospital, where his mother held a part-time job. He had suffered some horrific injuries, as later described in his biography, *For Spacious Skies*.

"The force of the impact had collapsed one lung. His ribs were cracked; he was bleeding internally …. His left foot was broken and mangled, his right knee sliced open and up through his quadriceps. He was bleeding profusely from face and scalp lacerations – he had, if fact, been neatly scalped. His entire scalp hung, inside out, off the back of his head."[8]

There is a popular misconception that the prominent scar on his forehead in later NASA portraits came from this incident, but he told the author that it actually resulted from another accident, many years before.

"I got a lot of scars from the crash, but I carried that particular scar into the accident. I was an infant. But my mother told me I somehow climbed out of my crib, and on my way to the floor I hit a radiator …. I was interested to note in later years that – my folks were pretty poor at that time – they did not have the money to go to a doctor to sew up this gash, which was almost an inch long, but it's the shape of one of your Australian footballs."[9]

Although he had been critically injured in the automobile accident, Carpenter realized he was incredibly lucky to still be alive, and his close brush with death changed his outlook on life. In 1949, following a long convalescence, he went back to university, where, two years after his accident, he met pretty Boulder co-ed Rene Price. As related in *For Spacious Skies*, "She was shelving textbooks at the Colorado Book Store when Scott walked in, scanning the store. She stepped up on her ladder, the better for him to see her. They had met once before, briefly – a fateful eye click in the lobby of the Boulder Theater, he in a black tux and bow tie, taking tickets, she with someone else."[10]

He stepped boldly up to the plate and invited her to a dance, and the rest was history. It proved to be a wonderful match, as described by Rene Carpenter in a 1959 interview for *Life* magazine.

"A long time ago, when Scott was a senior at the University of Colorado, he and I went out of our way to isolate ourselves because we knew it would bring us closer together. For our first home we picked a remote house in the mountains, seven miles from the nearest neighbor. Scott went out every day to chop wood from a pile of discarded telegraph poles, then hauled it in to feed our fireplace and kitchen stove. It was cold and primitive, but it was a wonderful first year of our marriage because it gave us a closeness we probably could not have achieved in softer surroundings. We still have it."[11]

FLIGHT TRAINING

With a new wife and associated responsibilities, Carpenter finally decided it was time to discuss his post-university job prospects with Rene. Flying remained a dream ambition, and he had heard that the Navy was recruiting engineering students through the Direct Procurement Program, which offered a commission as an officer, coupled with flight training. But it required a university degree. Rene supported her husband's decision, even though she knew that if he were successful it would mean that they would be frequently separated.

Much to his disappointment, Carpenter failed to gain his aeronautical degree from the University of Colorado, as he could not understand or pass a course on heat transfer. However, the Navy assumed in error that he had passed his degree and accepted him as a trainee aviator. He wisely said nothing and managed to remain in the service, reporting for training in October 1949.

The first six months of training at Pensacola Naval Air Station in Florida involved classroom studies encompassing such subjects as celestial navigation, radio procedures, and aircraft recognition. While he was completing his training and preparing for his first solo flight the Korean War began, and once again he was keen to enter the conflict. During his training, Scott and Rene had their first child: Marc Scott, who became known as "Scotty."

After mastering the basics of flying and moving on to aerobatics, gunnery, and night flying, Carpenter was ready for the next phase, carrier qualification, which involved landing practice – theoretically on the deck of an aircraft carrier, but actually in the relative safety of an outlined runway on the ground. Six such exercises completed his initial training, and he was ready to move on in his quest to become a naval aviator.

Next, Carpenter successfully passed through advanced aircraft training at Corpus Christi, Texas, during which time he had a difficult decision to make. As a naval aviator he could choose one of three non-changeable alternatives for his future career in the service. For many years he had wanted to become a hot-shot fighter pilot, but at the age of 26, with another baby now on the way, he had to choose between this most dangerous of options, or flying patrol planes, or perhaps even helicopters. Rene was finding it hard to understand why he was procrastinating on this, as his life's ambition had always been to serve his country as a fighter pilot. But he said he had looked at it responsibly and did not want to be an absentee father, or at worse leave her a widow and their children without a father. In the end – with some reluctance – he made his decision and took on the safer option of four-engine flying. Thus he completed his advanced training on airplanes such as the Consolidated PB4Y-2 Privateer heavy bomber and the Lockheed P2V Neptune. On 19 April 1951, a proud Rene was finally able to pin onto her husband's uniform the wings of gold of a naval aviator.

He subsequently spent three months in the Fleet Airborne Electronics Training School based in San Diego, California, before joining a Neptune P2V training unit at Whidbey Island, Washington, where he remained until October 1951. Sadly, during this time, he and Rene lost their second son Timmy, who died in his sleep of a virus infection at the age of just six months.

As Rene recalled, "Scott's faith in God and his own incredible strength brought us through that tragedy, and it probably made Scott more conscious of his role as a father than most men."[12]

TEST PILOT SCHOOL

In November 1951, still with a heavy heart over the loss of their baby boy, Carpenter was assigned to Patrol Squadron 6 (VP-6, "The Blue Sharks"), based at Barbers Point, Honolulu. Rene – pregnant once again – and two-year-old Scotty accompanied him to Hawaii, where they were housed in a garden apartment complex close to Waikiki Beach as the squadron prepared to embark for their forward base at NAS Atsugi in Japan.

Carpenter saw service in the Korean conflict, first as a navigator, with a chance of then being promoted to copilot and eventually plane pilot. He began by flying anti-submarine patrols with VP-6 in the lumbering multi-engine P2Vs, in addition to conducting mine-laying missions and shipping surveillance work in the Formosa Straits, the Yellow Sea, and the South China Sea. "These aircraft are chock-full of electronics and communications gear and spend most of their time flying over water on shipping surveillance," he would later record. "This was exactly what I did during the Korean War, as a matter of fact, while some of the other fellows were shooting down MiGs or plastering the front line with napalm."[13]

In January 1952 the squadron returned to Hawaii. Two months later Rene gave birth to another son they named Jay, and then the squadron members and their families moved into a series of Quonset huts on Honolulu's Ewa Beach. All too soon it was time for VP-6 to undertake the squadron's second wartime deployment, this time based at NAS Kodiak in Alaska. By now, having moved up the chain to pilot, and with an excellent service record under his belt, Carpenter had turned his eyes to a whole new goal – to enter the U.S. Naval Test Pilot School at NAS Patuxent River, Maryland.

Carpenter chats with Jane Howard, the wife of VP-6 C.O. Guy Howard at Barbers Point, Hawaii, in 1952. (Photo: Patrol Squadron Six VP-6 Association)

VP-6 would return after New Year's in 1953 and later completed their third deployment, this time to NAS Agana, on the island of Guam. By now he had been further promoted to patrol plane commander.

The conflict in Korea finally ended on 27 July 1953 with the signing of the Korean Armistice Agreement and the squadron returned to Barbers Point. While continuing his service there as a plane commander, Carpenter endured one incident that he felt gave him the confidence later on that he would succeed as an astronaut.

"We were flying out of Barber's Point in Hawaii, and we were supposed to go about 250 miles out on a certain leg, then fly back to get a navigation check, then go out again. It was about an eight-hour mission and the night was very black. I had finished the first two legs and the next one took us out over a really dark sea. There were thunderstorms in the area, and I could see flashes of lightning playing around. I knew there was going to be some turbulence, and it was not exactly pleasant to turn my tail on the friendly lights of Honolulu and keep on going. Looking back on it now, it sounds a bit silly. But it takes little moments like that to build up a person's tolerance of fear and his ability to face the unknown."[14]

Scott and Rene in Hawaii with sons Scotty and Jay. (Photo: Carnegie Branch Library for Local History)

In May 1954, Carpenter was promoted to full lieutenant. That was not the only welcome news he received; squadron commander Guy Howard said he was recommending him for Test Pilot School. His appointment to the school was later confirmed, and Scott Carpenter was ready to enter a whole new phase in his life, finally fulfilling his boyhood dream of flying the fastest, newest and sleekest airplanes in the world. On the home front, Rene also confirmed that they were having another baby.

He would spend the next three years at Patuxent River. For the first six months he was a student in Test Pilot School as one of the 30-strong Class 13, who named themselves the "Thirsty Thirteen." It would prove an exhausting, intense time. First up, the students had to undergo refresher courses involving such disciplines as algebra, trigonometry, calculus, and mechanics. As well, each pilot ran a flight schedule, carrying out practice flight tests nearly every day in familiar aircraft, then preparing detailed flight reports, linking the classroom theory with actual flight tests. They were not being taught how to fly, as they were already talented pilots; what they were being taught to do was to write reports on the capabilities of aircraft.

The aircraft types they were assigned during this time were an exciting mix, ranging from relatively simple propeller-driven trainers such as the North American T-28B Trojan and the Beech SNB-5 Navigator, through to single engine fighter/attack aircraft including the Grumman F8F-1 Bearcat, the Douglas AD-3 Skyraider, and the Grumman F9F-2 Panther. They also flew carrier-based fighter jets such as the Grumman F9F-6 Cougar and the swept-wing North American FJ-2 Fury. As the U.S. Navy was not confined to piston- or jet-powered attack aircraft, it also flew larger, purpose-built airplanes such as the twin-engine anti-submarine warfare Grumman XS2F-1 Tracker, the Grumman UF-1 Albatross air-sea rescue flying boat, and the Consolidated P4Y-2 Privateer four-engine patrol bomber.[15] It was a wonderfully diverse range of aircraft types and gave the Patuxent aviators a broad operating experience which would serve them well in their future service careers.

According to Carpenter, "I spent three years at Patuxent, where my daughters, Kristen and Candace, were born. I studied hard and flew hard, graduating in the top third of my class. For the first time in my life I had access to every conceivable airplane available to free world pilots and flew every one I could."[16]

After graduating from test pilot school, Carpenter would continue at the Naval Test Center until 1957, serving as a test pilot in the Electronics Test Division, conducting flight-test projects on advanced aircraft such as the Douglas AD-3 jet bomber, the Grumman F9F Panther and the Grumman F11F Tiger jet fighters. It was a nervous period for his wife, Rene.

"I still remember how I felt sometimes when he was a test pilot at Patuxent River. If he was flying on a project and did not come home by six o'clock, I just knew I was a widow. I remember wondering once how I was going to greet the chaplain when he came to the door. That sounds morbid, I know, but it creeps in every now and then. I can't keep it out. I don't think any woman could shut it off completely."[17]

NAVAL INTELLIGENCE

In October 1957, now 32 years old, Carpenter had just completed a year's training in electronic intelligence (ELINT) at the Navy's Postgraduate School in Monterey, California when the Navy assigned him to Air Intelligence School in Washington, D.C. Following this, he could expect to serve a mandatory three-year tour of sea duty as an intelligence officer. The prospect of this, and the long separation from his family, was not something he relished.

Carpenter's travel orders to Washington meant a cross-country trip, and he decided it would be a great and bonding adventure if Scotty and Jay joined him part of the way on a drive-and-camp trip across to Boulder. Rene, having flown ahead with Kristen and Candace, would meet them there as planned. He would then continue by himself to Virginia, ready to report to Air Intelligence School by 31 October.

He recalled camping with the boys one clear night in Nebraska, with all three lying back on the ground and keeping a close watch on the brilliant, starry night sky. He had heard news on the car radio of the Soviet Union launching a satellite called *Sputnik* into Earth orbit, and as they scanned the skies in expectation he told his sons how this significant event would herald in a new age of human exploration. Suddenly they caught sight of the upper stage of *Sputnik*'s carrier rocket gliding serenely across the night sky. (The actual satellite, flying nearby, was too small to be seen.) The sighting caused great excitement, but as they talked about space and what wonders lay beyond the Earth, Carpenter had no inkling just how much that simple, beach-ball sized satellite would help forge his own future.

The launch of *Sputnik* caused an immediate uproar across the United States, with people concerned that the Soviet Union would soon have the capability and high-ground means to rain nuclear warheads onto the United States from space. Alarmed by the possible threat to national security and a perceived lack of technological parity with the Soviet Union, the United States Congress demanded immediate and swift action. President Dwight Eisenhower, while never an advocate of throwing money at anything as fanciful as a space program, quickly came to the uneasy realization that the United States would have to play catch-up and develop the means to launch Americans into space ahead of the Russians. What became known as the Space Race had begun.

PROJECT MERCURY

Following months of consultation with the nation's space experts and his own Science Advisory Committee, President Eisenhower signed the National Aeronautics and Space Act on 29 July 1958, which led to the creation of a civilian space agency, the National Aeronautics and Space Administration, better known by the acronym NASA.

Acting quickly, NASA had soon set in motion a program to select the nation's first spacefarers, who would be called "astronauts." After a great deal of to-and-fro debate among experts as to what sort of candidates were best suited for space flight, an impatient President finally intervened. The candidates, he decreed, *had* to come from the ranks of jet-qualified military test pilots. Not only were they fully trained pilots used to rating the performance of different and innovative aircraft, but their service and medical records were complete and immediately available for inspection. Once this was agreed, NASA put together a working group to establish the basic criteria for applicants and to subsequently identify and select the final candidates. It was felt that around six pilots would be needed for the initial flight phase of what was known as Project Mercury, so they set down specific guidelines, detailing a maximum age, flight experience, height, education, intelligence, temperament, and physical ability.

By this time Scott Carpenter had completed his intelligence work for the Navy in Washington, which he described as a good and worthwhile experience. "I learned quite a bit about photo reconnaissance and received some more training in navigation as well as electronics and communications. I think all of this helped me some when NASA looked me over for Project Mercury."[18]

He had accumulated around 2,800 flying hours, 300 of which was accrued operating jet aircraft, but after his interesting work experience with Navy intelligence Carpenter now faced the daunting prospect of following naval regulations. These stipulated that all naval men had to spend certain periods at sea, and his assignment in August 1958 proved to be a non-flying desk job as intelligence officer aboard the anti-submarine aircraft carrier USS *Hornet* (CVS-12). Although he knew it was something he had to face in his chosen career, he was far from happy with this appointment. He knew it would not only keep him away from flying, but he would not see his family for long stretches of time. Also, coming up on the 10-year mark in his service, he had failed to gain an expected promotion. Deeply discouraged, he began to contemplate leaving the Navy and taking on a less dramatic life as a civilian. For the moment, though, he followed orders and made his way to Long Beach, California, where the *Hornet* was berthed.

Months later he received a message – ominously marked Top Secret – that would dramatically alter the course of his life.

"In January 1959, just as we began our sea trials out of Coronado, I received mysterious orders from the Chief of Naval Operations, the legendary Admiral Arleigh A. Burke. I was to report to the Pentagon on February 1 without discussing my orders or speculating about them. There was no explanation. Nothing. Then again, I was used to orders; I had been following them for more than a decade. I had little inkling about how momentous my orders were for the country, for my career in the Navy, for my family, or for me."[19]

After showing the ship's captain his orders, Carpenter was flown off the carrier and caught a train home, where he and Rene pondered what this secret program might be. He assumed it might have something to do with his intelligence work for the Navy. While en route to Washington, Carpenter happened to pick up the latest copy of *Time* magazine at Los Angeles airport and read with growing interest that the National Aeronautics and Space Administration had selected 110 military test pilots as potential candidates for the Project Mercury man-in-space program. He realized this could be the reason behind his secret orders, and he was certainly interested. "I was not too happy with what I was doing and I was eager to make a change," he once reflected.[20]

It was planned that the 110 men would be given extensive briefings by service personnel and NASA over three separate dates. The first two briefing sessions involved 69 candidates from the Navy, Air Force and Marines. Carpenter's Navy group received their initial briefing from Vice Admiral Robert Pirie, who gave the men an outline of the project and assured them the Navy fully supported the NASA program. It was stressed that any man who volunteered would be regarded as "on loan" to NASA, and their normal professional progress and promotions would continue. However, should any man wish to decline the opportunity it would not be notated on their record and they could return without prejudice to their present duties.

The service groups then combined for a lengthy briefing given by Charles Donlan, a senior NASA engineer; Warren North, a NASA test pilot and engineer; and Lt. Robert Voas, a Navy psychologist. They outlined the NASA organization and presented an overview of Project Mercury. Several pilots, mindful of giving up current duties and opportunities within their service to take on a completely different and uncertain role, declined. Eventually, when an unexpected high ratio of 80 percent of the first two groups volunteered after hearing what was involved, the third session of 41 lesser-qualified military pilots was cancelled.

As Carpenter recalled, "The people at NASA were looking for pilots with different kinds of talents so that the astronauts on the team would complement each other. They apparently needed someone who understood a little more about communications and navigation than the average pilot would have learned, and I could help out here. I still felt slightly self-conscious about my flying, however. Most of the other fellows I was competing with had a total of more than one or two thousand hours in jets. I had only 2,800 hours of any kind of flying, and only 300 of this was jet time. I suppose this should not have concerned me, but it did. I was worried that I had not had a real chance to prove myself. My nerves and reactions were all right, however, and my record was good."[21]

After the briefing, Rene received a phone call from her husband, in which they discussed the program and all its implications. She was excited and supportive. He promised that he would send her a letter; which he did.

"We had our first briefing today," Scott wrote to Rene from Washington on 6 February. "The competition will be tremendous, but I know if there is anyone who can do a job like this, I can do it, and if I am given the chance, I must take it. I may sound a bit heady about the whole thing, but the project gives me the opportunity to use all my capabilities and interests at once. It even exceeds my boyhood dream of roughing it in some strange country. In spite of all other considerations, here flickers something that some men search their entire lives for – it may mean my fulfillment not as your husband, head of family, but as a man."[22]

Following a series of probing interviews and medical assessments, the number of suitable candidates was reduced to 32. These men would undergo several weeks of intense physical and psychiatric examinations, brutal stress tests, and psychological evaluations, firstly at the Lovelace Clinic in Albuquerque, New Mexico, and then the Wright Air Development Center in Dayton, Ohio. There was only one problem for Scott Carpenter – he was at sea and knew nothing of a letter from NASA that came for him, saying, 'If you wish to continue in the program, contact this office by Monday.' Ever the good wife, Rene got on the phone, called NASA headquarters, and volunteered her husband for the assignment. He made it.

The 32 finalists were subsequently prodded, thumped, punctured, and X-rayed. The test results were collated and extensively debated by the Space Task Group's selection committee over the following two weeks. On 1 April the final list contained just seven names – seven men who would become known as America's Mercury astronauts. The name of Scott Carpenter was first alphabetically on that list.

"I think one of the main reasons I made the team – aside from the knowledge of navigation and communications that I could bring to it – was that I managed to do quite well on the physical tests. I was the sort of physical specimen they were looking for, I guess, and I was in good shape. The doctors confided to me when they were totting up the scores that I had broken five of their records. The treadmill was one of these. This was a walk that moved uphill underneath you at a constant speed, and you had to keep pace with it with your legs. It was like climbing a mountain that kept getting steeper and steeper, and it was tiring, to say the least. But it was immensely satisfying to me to know that I kept going longer than anyone in my group."[23]

He also broke records in respiration tests and in the altitude chamber. The physicians at Wright-Patterson and the Lovelace Center were highly impressed with his level of fitness and stamina. There was another test, which featured prominently in the 1983 movie, *The Right Stuff*.

"Another record fell when the doctors asked me to blow into a tube that held a column of mercury. The purpose of this was to see how long I could keep blowing before I ran out of wind. The record up until then was 94 seconds. I thought to myself, I'll just count up to one hundred as fast as I think seconds ought to go and I'll try to break this record, too. I counted too slowly, however, and when I finally had to give up I found I had blown into the tube for 171 seconds without taking a breath. This little victory pleased me very much. The doctors were jumping up and down. I was next to the last in our group to take this test.

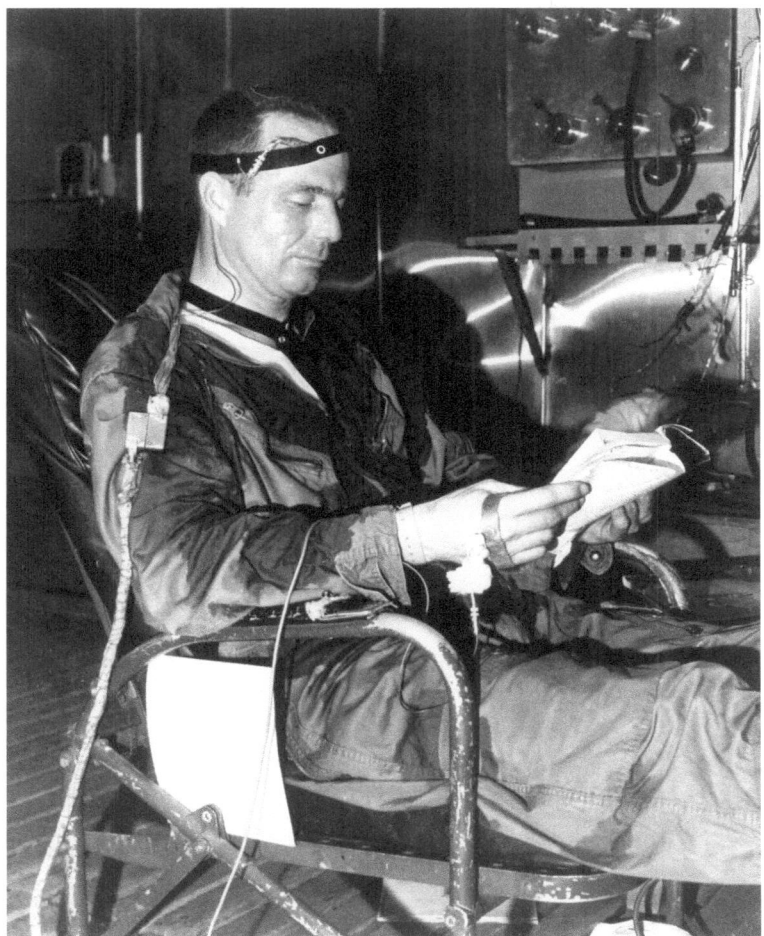

As part of the tests carried out at the Wright Air Development Center, the astronaut candidates were subjected to high temperatures in a heat chamber, with physicians checking a candidate's tolerance to heat and their ability to function under stressful heat loads. In completing this test, Carpenter had to endure 130°F for nearly two hours. (Photo: U.S. Air Force)

Then John Glenn came along and managed to blow into the tube for 151 seconds. John did well all through these tests, too. He told me one day that he had overheard the doctors say the two of us were doing so well they were going to tell Washington about us."[24]

"While I was waiting to see if I had been picked for the team, I had to go back to my ship once more," Carpenter recalled. "She was in port at Long Beach now, but she was due to sail almost immediately for another six-month cruise, and I was in for a fairly close call.

"The day before the ship left, I got word from the officer of the day to call someone in Newport News. It was Mr. Charles Donlan, the associate director of Project Mercury who was in charge of putting together the team. Mr. Donlan said, 'We'd like to have you aboard.' I said, 'Fine, I'll be your hardest worker.' Then I went back to the ship. I was elated."[25]

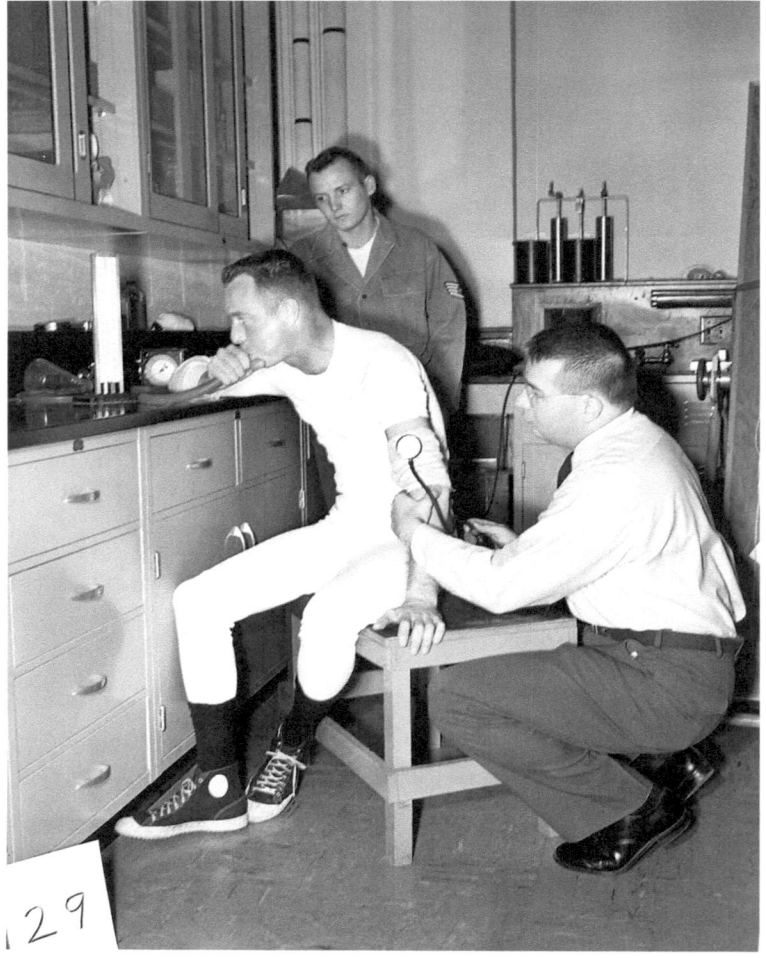

Holding a column of mercury as a breath endurance test. (Photo: U.S. Air Force)

One person who did not share Carpenter's elation was the ship's captain, who was furious that his intelligence officer was leaving the carrier yet again, this time for good.

"Captain Marshall White was disappointed in me," Carpenter explained. "I was their intelligence officer, and I had to leave the ship three times, for a week at a time, while we were training for deployment. He was getting very angry at me leaving, and it took the CNO [Chief of Naval Operations] on the telephone to call Marshall White and say, 'I know you need an intelligence officer for your next deployment, but right now the nation needs yours, and we will send you another.' Before I left the ship, after the telephone call saying that I had been selected, everyone else was walking aboard with their gear. I was walking off, and he finally in reasonable and understandable impatience – remember, this is a trained, dedicated naval officer – said, 'Would you mind telling me what this is all about?' Of course, I was bound to secrecy by these orders. I said to this fine navy captain, 'I am not supposed to tell you, captain, but I am going to Washington to report to NASA, and I am going to ride the nosecone of a rocket around the Earth.'

"Try to think of his answer!"[26]

SEVEN FOR SPACE

The selected men were introduced to the public on 9 April 1959 during a packed press conference at the Dolley Madison House in Washington, D.C. The seven Project Mercury astronauts were named – in alphabetical order – as Lt. M. Scott Carpenter (USN); Capt. Leroy G. Cooper (USAF); Lt. Col. John H. Glenn (USMC); Capt. Virgil I. Grissom (USAF); Lt. Cmdr. Walter M. Schirra (USN); Lt. Cmdr. Alan B. Shepard (USN); and Capt. Donald K. Slayton (USAF).

The seven Mercury astronauts were introduced at a Washington press conference. From left: Donald ("Deke") Slayton, Alan Shepard, Wally Schirra, Gus Grissom, John Glenn, Gordon Cooper and Scott Carpenter. When asked, "Who wants to be the first man launched into space?" all seven raised their hands. Schirra and Glenn gleefully raised both hands. (Photo: NASA)

Until special living and working quarters were established at Cape Canaveral, the newly selected astronauts were to report for training activities under the auspices of NASA's Space Task Group, based at Langley Field on the southern tip of Virginia. Here they would share a small, non-partitioned office in a two-story building belonging to the Air Force.

Knowing that he was now assigned to Langley Field, Scott and Rene had to pack their belongings and move the family out of their home in Long Beach, California. In the astronaut book, *We Seven*, he tells an amusing story about being at home one day, busy packing his belongings, when the landlord brought a lady over to show her through the house.

"He had told her who I was. She came into the room where I was working and asked me if I was really one of those new spacemen. I told her I was, and she said, 'You're a nut!' Then she walked off to look at the kitchen."[27]

The Project Mercury astronauts standing beside a Convair F106-B aircraft. Lined up in alphabetical order from left: Scott Carpenter, Gordon Cooper, John Glenn, Gus Grissom, Wally Schirra, Alan Shepard and Deke Slayton. (Photo: NASA)

EARLY TRAINING AT LANGLEY

When the seven astronauts were not in their Langley office they were out visiting different companies, factories, and installations associated with the space program to get a feel for the work ahead of them, or training with equipment at facilities not available to them at Langley. While five of the men and their families took up temporary residence near to Langley Field, Scott Carpenter and Gordon Cooper chose to live on the base with their wives and children.

The astronauts began their training for space flight with the Langley engineers, who educated them in graduate-level space sciences courses including the fields of atmospheric re-entry physics, astronomy, celestial mechanics, and navigation. Meanwhile the STG staff introduced the astronauts to numerous space flight simulation systems and techniques to help familiarize them with the Mercury capsule and evaluate their effectiveness with the capsule's control systems. A closed-loop analog simulator became the basis for several of these simulated flights into space. It was simply equipped, being initially furnished with a chair which had a sidearm controller and rudder pedal. Later on it was refitted with a three-axis controller and a molded couch individually manufactured for each astronaut.

The training at Langley also included a regimen of physical exercise and scuba-diving operations designed to simulate weightlessness and the types of sensory disorientation that they might experience during re-entry from space. In Langley's large hydrodynamics tank and in the Back River behind the Langley East Area, the Mercury astronauts also learned egress techniques from the space capsule as it floated in the water.[28]

The astronauts would initially maintain their flying proficiency in NASA's Convair F106-B Delta Dart jet aircraft. (Photo: NASA)

Scott Carpenter working in a raft during capsule egress exercises. (Photo: NASA)

Egress techniques were also practiced on non-functional capsules on Back River. (Photos: NASA)

All too soon, the astronauts found that it was becoming increasingly unmanageable as a group to visit every company, contractor, or subcontractor, and digest all there was to know about every facet of the Mercury program. This included the ongoing development of the spacecraft and its many intricate systems, the booster rockets, and hundreds of ancillary tasks and procedures associated with sending humans into space. There were also stacks of complex technical manuals to study. And all this was in addition to their arduous astronaut training program. It was decided to split the work up between them, and give each man an area of responsibility. This way they could each concentrate their efforts on a specialized field of activity, then report back to the others at regular meetings which they whimsically referred to as "séances."

In this way, Gordon Cooper took on responsibility for the Redstone rocket system, including such elements as configuration, trajectory, aerodynamics, countdown, and flight procedures, and Gus Grissom worked alongside McDonnell engineers on the hand control and autopilot systems. Wally Schirra's area of specialty was the Mercury pressure suit and the environmental systems, while Alan Shepard worked on Mercury recovery operations. Deke Slayton's involvement was with the Atlas rocket and flight procedures, and John Glenn's specialized area was the layout of the Mercury spacecraft's cabin systems and instrumentation.

Due to his prior and extensive experience in the field, Scott Carpenter was handed the task of communications and navigation systems. "I'd flown some Navy airplanes that had equipment similar to our [capsule's] periscope, which was our primary visual navigator. We couldn't navigate in the classic sense … [but] could keep track of our progress through the periscope. That was my area of expertise, but we all crossed over a great deal into all areas of the spacecraft design."[29]

One of the main contractors the astronauts regularly visited was the General Dynamics plant in San Diego, where the Atlas boosters were manufactured. (Photo: NASA)

The astronauts leaving Stead Air Force Base for desert training. From front: Cooper, Carpenter, Glenn, Slayton and Grissom. (Photo: Reno Gazette-Journal)

Survival training in the Nevada Desert. (Photo: NASA)

Carpenter involved in sea retrieval exercises in Virginia's Back River. (Photos: NASA)

As well as their regular meetings to brief each other on their areas of specialty, all seven astronauts traveled to the Morehead planetarium in North Carolina to learn about celestial navigation. Then, on the off-chance that their spacecraft ended up landing well away from the planned splashdown site, they were sent on a trip to Fallon, Nevada, for desert survival training. During this exercise they were taught how to make protective clothing out of their parachute cloth. The cloth was also used to make small, crude tents to shelter them from the blazing desert Sun. They later took part in ocean egress exercises off Pensacola in the Gulf of Mexico, using makeshift Mercury training capsules while bobbing around dressed in full pressure suits.

Most of the astronauts agreed that one particularly vital training device was the giant centrifuge located at the Navy Aviation Medical Acceleration Laboratory in Johnsville, Pennsylvania, which was used to accustom them to heavy g loads. The sole exception was Wally Schirra, who stated that in his opinion it was little more than "a rough ride" that was no use at all.

Carpenter endures heavy g-loads whilst training in the Johnsville centrifuge. (Photo: NASA)

Spending time on the Procedures Trainer at Langley Field. (Photo: NASA)

Another device was the flight procedures trainer, a complicated device which consisted of a mock-up of the Mercury capsule, with all of its systems connected to control panels and computers. In this way the trainee could "fly" a mission, simulate emergency situations, and have their proficiency tested, monitored, and expanded – all within the trainer.

In a conversation with former NASA Public Affairs officer Ed Buckbee in 2005, Scott Carpenter revealed his thoughts on the dynamics involved in the space flight training the astronauts went through.

"Because there were so many unknowns about how the human organism would respond to these new stresses, the spacecraft development itself sort of lagged. We probably were more thoroughly investigated in psychological and physiological ways than any other group of men, ever. The thing that amazed me was that this human organism is so miraculously designed as to be adaptable to these stresses. The human body is a marvelous machine. On the centrifuge, for instance, we had no evolutionary experience with high acceleration. But lo and behold, there is almost an automatic defense to that high acceleration. You don't need training, but practice helps. You know automatically how to combat it and I think that is an amazing thing. The human body is very well adapted to space flight and the rigors with which it has no evolutionary experience. It's a marvelous thing."[30]

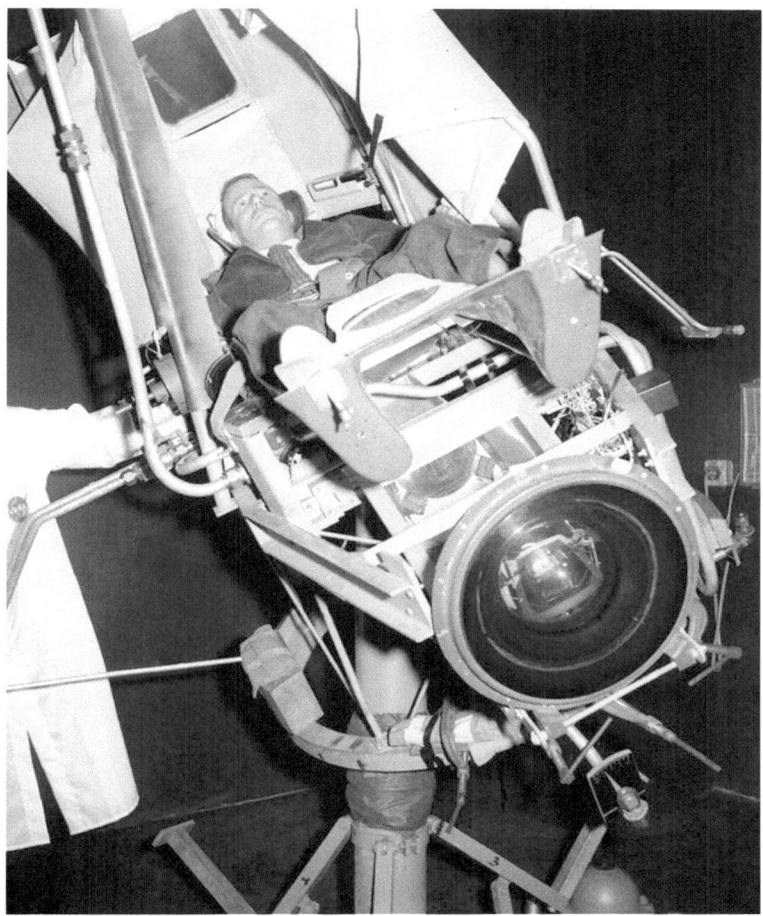

Carpenter demonstrates the use of the ALFA trainer. It furnished practice in using the manual controls that keep the Mercury capsule in proper alignment during flight. The mirror in the lower foreground is part of the optical system which gave the astronaut a visual check on the relationship of the capsule and the Earth during simulated flight. (Photo: NASA)

Yet another crucial training device for the astronauts was the Air Lubricated Free Attitude (or ALFA) trainer, which consisted of a couch so perfectly balanced on a cushion of compressed air that it was practically frictionless. Moving freely on three axes, it provided the means for the future astronauts to practice keeping a space capsule at the proper orbital attitude. It was practicing on this particular device that would greatly assist Scott Carpenter on his later orbital space flight.

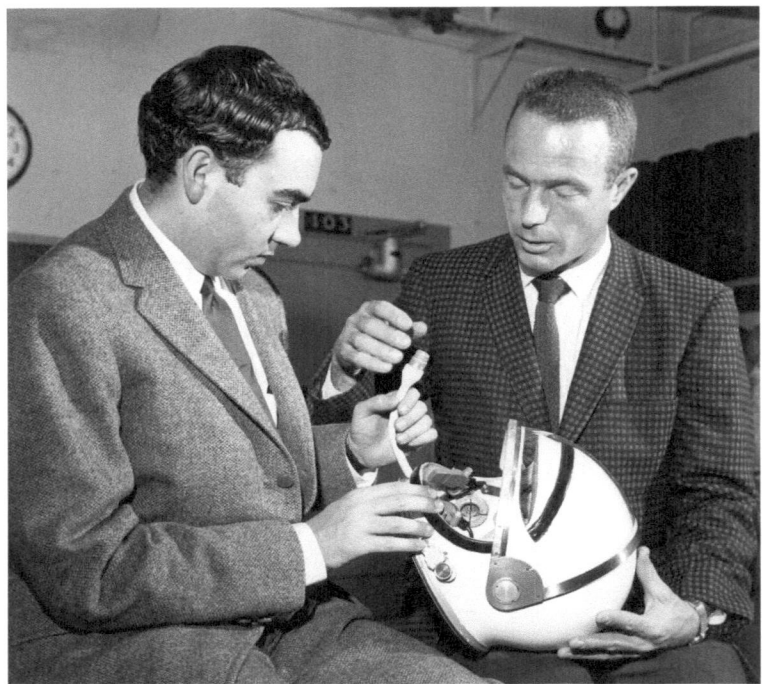

Scott Carpenter discussing his Mercury space helmet with a technician at the B. F. Goodrich plant. (Photo: NASA)

But as the seven astronauts continued their training, the news emanating from Cape Canaveral seemed to be routinely dominated by vivid descriptions of one highly spectacular rocket explosion after another.

On the evening of 18 May, just three weeks into their training, the men had gathered at the Cape to collectively watch their first missile launch. The Atlas 7D rocket looked spectacular on the East Test Range Pad 14, gleaming in the glare of huge searchlights. The clock slowly wound down, and then the rocket's three powerful engines ignited and the vehicle surged off the pad. Just over a minute after liftoff it suddenly exploded in a massive fireball in the night sky. Apart from some shocked exclamations, the seven men stood in stunned silence. Was this really the rocket they wanted to ride into orbit? Even Wally Schirra, famed for his puns, had nothing to say. It was left to Alan Shepard, who calmly shrugged.

"Well," he said, "I'm glad they got *that* out of the way."[31]

18 May 1959, and a gleaming Atlas 7D sits on the launch pad at Cape Canaveral. Shortly after lifting off it blew apart in a massive explosion. (Photo: NASA)

REFERENCES

1. The Rehoboth Carpenter Family, Wikipedia entry, available at *http://en.wikipedia.org/wiki/Rehoboth_Carpenter_family*
2. Woody Paige, *Mile High Sports* magazine article, "Woody Paige interviews Scott Carpenter." Online at *http://milehighsports.com/2013/10/10/magazine-woody-paige-interviews-scott-carpenter*
3. Loudon Wainwright, "Comes a Quiet Man to Ride Aurora 7," *Life* magazine, issue Vol. 52, No. 20, 18 May 1962
4. *Ibid*
5. Michael Ryan, "The Astronauts Memorialized in Right Stuff Have Gone Surprisingly Different Ways," *People* magazine, issue Vol. 20, No. 18, 31 October 1983
6. Eugene (Oregon) *Register-Guard* newspaper, article, "Mrs. Florence Carpenter 'Intensely Proud' of her Son," issue 25 May 1962, pg. 6A

7. Francis French and Colin Burgess, *Into That Silent Sea: Trailblazers of the Space Era, 1961–1965*, University of Nebraska Press, Lincoln, NE, 2007
8. Scott Carpenter and Kris Stoever, *For Spacious Skies: The Uncommon Journey of a Mercury Astronaut*, Harcourt, Orlando, FL, 2002
9. Scott Carpenter telephone interview with Colin Burgess, 18 December 2002
10. Scott Carpenter and Kris Stoever, *For Spacious Skies: The Uncommon Journey of a Mercury Astronaut*, Harcourt, Orlando, FL, 2002
11. Rene Carpenter, "There Are No Dark Feelings," *Life* magazine, 21 Sept 1959, pp. 146–148
12. *Ibid*
13. Carpenter, S., Cooper, Jr. L, Glenn, Jr., J., Grissom, V., Schirra, Jr., W., Shepard, Jr., A., and Slayton, D., *We Seven*, Simon and Schuster Inc., New York, NY, 1962, pg. 50
14. *Ibid*, pg. 52
15. *United States Naval Test Pilot School, 1945 to 1983*, Fishergate Publishing Company, Inc., Annapolis, MD, 1983
16. Scott Carpenter, article, "Test Pilot and Naval Aviator." Available online at *http://www.scottcarpenter.com/test_pilot.htm*
17. Rene Carpenter, "There Are No Dark Feelings," *Life* magazine, 21 Sept 1959, pp. 146–148
18. Carpenter, S., Cooper, Jr. L, Glenn, Jr., J., Grissom, V., Schirra, Jr., W., Shepard, Jr., A., and Slayton, D., *We Seven*, Simon and Schuster Inc., New York, NY, 1962
19. Scott Carpenter, article, "Test Pilot and Naval Aviator." Available online at *http://www.scottcarpenter.com/test_pilot.htm*
20. Carpenter, S., Cooper, Jr. L, Glenn, Jr., J., Grissom, V., Schirra, Jr., W., Shepard, Jr., A., and Slayton, D., *We Seven*, Simon and Schuster Inc., New York, NY, 1962
21. *Ibid*, pg. 50
22. Rene Carpenter, "There Are No Dark Feelings," *Life* magazine, 21 Sept 1959, pp. 146–148
23. Carpenter, S., Cooper, Jr. L, Glenn, Jr., J., Grissom, V., Schirra, Jr., W., Shepard, Jr., A., and Slayton, D., *We Seven*, Simon and Schuster Inc., New York, NY, 1962, pg. 56
24. *Ibid*, pg. 57
25. *Ibid*, pg. 54
26. Scott Carpenter talk at the Reuben H. Fleet Science Center, San Diego, CA, 20 January 2003, hosted and transcribed by Francis French
27. Carpenter, S., Cooper, Jr. L, Glenn, Jr., J., Grissom, V., Schirra, Jr., W., Shepard, Jr., A., and Slayton, D., *We Seven*, Simon and Schuster Inc., New York, NY, pg. 59
28. "Langley's Role in Project Mercury," (FS-1996-04-29-LaRC), NASA Langley Research Center, Office of Public Affairs, April 1996
29. Ed Buckbee and Wally Schirra, *The Real Space Cowboys*, Apogee Books, Ontario, Canada, 2005
30. *Ibid*
31. John Glenn with Nick Taylor, *John Glenn: A Memoir*, Bantam Books, New York, NY, 1999

3

Mission planning for MA-7

As a public affairs officer for NASA at the Marshall Space Flight Center, Ed Buckbee got to spend a lot of time with Scott Carpenter whenever the astronaut visited Huntsville to meet with rocket designer Wernher von Braun. It always struck him how different Carpenter was to his fellow astronauts.

"When Scott visited von Braun the discussion was more about what's out there in the solar system and how far can we go with our technology. He asked von Braun, 'What can we do during our flights to help you build a better rocket?' Scott's interest was more in having a bigger window in the spacecraft so he could observe better and add some science experiments. He did not have the interest as other Mercury astronauts had of more control and bigger thrusters aboard the spacecraft. His focus was more on studying the effects of weightlessness on the human body and conducting science experiments."[1]

LOADING UP WITH EXPERIMENTS

Many questions relating to orbital space flight had been answered by John Glenn's MA-6 mission, and this – as expected – aided in setting the flight program for Scott Carpenter. On 28 February, a week after his history-making three-orbit flight aboard *Friendship 7*, Glenn appeared before the U.S. Senate Aeronautical and Space Sciences Committee along with Alan Shepard and Gus Grissom. They told the committee that as a result of Glenn's flight, American space scientists were planning to have astronauts on future flights steer by the stars. Glenn told the senators that he saw stars when he was on the dayside as well as the nightside of Earth. Shepard said that this being the case, the astronauts intended using stars as fixes for navigation. They also reported that Glenn's flight had proved the value of having a pilot rather than only automatic equipment in human-tended spacecraft. "A human being can work more effectively, pound per pound, than any mechanical computer," Shepard declared.[2]

© Springer International Publishing Switzerland 2016
C. Burgess, *Aurora 7*, Springer Praxis Books, DOI 10.1007/978-3-319-20439-0_3

Aurora 7 goes through a weight and balance check in Hangar S. (Photo: NASA)

Scott Carpenter practices insertion techniques while *Aurora 7* is inside Hangar S. (Photos: NASA)

In the light of the success of the MA-6 mission, it was decided that the MA-7 flight plan should permit far more pilot control. Yaw and roll maneuvers were therefore incorporated, allowing Carpenter to fully observe each sunrise, and maneuver *Aurora 7* in order to research the potential use of day and night horizons, landmarks, and stars as navigation aids. This was perfectly fine by him; in his autobiographical book *For Spacious Skies*, written in the third person, he states, "John [Glenn] had proved a man could handle the machine under difficult circumstances. To Scott it was obvious that the next astronaut in space had the freedom, amounting to a responsibility, to observe, measure, and study events *outside* the capsule."[3]

One innovation planned for Carpenter's voyage involved a period of inverted flight (head toward Earth) to determine the effect of Earth-up and sky-down on pilot orientation. Flight planners recognized the need for perceptual reorientation in space flight as well as for the motor skills that had been demonstrated so well by Glenn. It was therefore felt that the next scheduled Mercury mission should incorporate as many scientific experiments as possible.[4]

Homer E. Newell, Jr., Director of NASA's Office of Space Sciences since 1 November 1961 (and essentially third in charge at NASA), realized that John Glenn had been able to respond to many of the scientific astronomical observation requests made prior to his flight. Newell therefore decided that the direction of the scientific portion of the manned space flight program should become the responsibility of a formal investigative committee. He subsequently appointed Dr. Jocelyn Gill to head up what became known as the Ad Hoc Committee on Scientific Tasks and Training for Man-in-Space, consisting of representatives from the various scientific disciplines. Dr. Gill, who held a doctor of philosophy degree from Yale University, was a former laboratory assistant and instructor of astronomy at Smith, Mount Holyoke, Wellesley and Arizona State Colleges before joining NASA in October 1961 as a staff scientist in astronomy and astrophysics.

Homer E. Newell, Jr. (Photo: NASA)

On 16 March 1962, two days after receiving the assignment, Dr. Gill organized a meeting of committee members in order to review past activities, to present a preliminary analysis of the scientific results of the MA-6 mission, to outline future objectives, and to set out tasks and goals for the next meeting. One of the principal aims of the committee was to devise a curriculum which would provide astronauts with the best available sources of information regarding any spatial phenomena they might observe on their respective missions. They also began to compile a suggested list of experiments for MA-7 which was relayed to the Manned Spacecraft Center.[5]

In turn, Kenneth ("Kenny") Kleinknecht, head of the MSC Mercury Project Office, gave Lewis R. Fisher, chairman of the Mercury Scientific Experiments Panel, responsibility for administering and arranging the experiments suggested by the ad hoc committee. Fisher and his associates were tasked with reviewing all proposed experiments from an engineering feasibility standpoint in terms of their scientific value, relative priority, and suitability for orbital flight.[6]

Fisher's panel assembled for the first time at Cape Canaveral on 24 April 1962, and collectively decided to prioritize five suggested experiments:

1) Releasing a multi-colored balloon that would remain tethered to the capsule (proposed by the Langley Research Center).
2) Observing the behavior of liquid in a weightless state inside a closed glass bottle (proposed by the Lewis Research Center).
3) Using a special light meter to determine the visibility of a ground flare (proposed by the Massachusetts Institute of Technology Instrumentation Laboratory).
4) Making weather photographs with hand-held cameras (proposed by the Weather Bureau).
5) Studying the airglow layer (proposed by the Goddard Space Flight Center), for which Carpenter would receive special training.

A MULTI-COLORED BALLOON

The first test on the list to be conducted on the MA-7 mission was an experiment to determine how colors reflect in space, using a balloon towed behind the spacecraft and what was referred to as "confetti." *Aurora 7* was therefore equipped with a system to deploy a tethered balloon during the mission's orbital phase. In anticipation of this experiment being approved for flight, work on the project had begun at NASA's Langley Research Center in January 1962, and a rigorous qualification test was followed by delivery of the packaged unit to Cape Canaveral on 13 March. With the experiment confirmed for flight on MA-7 the unit was then installed in *Aurora 7*.

The test apparatus consisted of a 30-inch, Mylar-aluminum sphere, inflated by an attached 900 psi nitrogen bottle. The balloon was divided equally into five segments, or lunes. The colors of these surfaces were orange, white, silver (aluminum), yellow and phosphorescent, which glows at night. The balloon would be tethered to the capsule with a 100-foot nylon line and an eight-foot strip of .005 gauge aluminum, which would act as a form of shock absorber. A small metal beam, instrumented with a strain gauge, provided the means of measuring drag. Electric squibs would activate the spring-loaded deployment and line-cutting mechanisms.

Another objective would be provided by the simultaneous dispersion of a cluster of small particles, or "confetti." It was intended that the visual effects and the behavior of these known objects could be closely studied by the astronaut. The 0.25-inch Mylar discs were placed between the folds of the balloon to provide the small particle cluster. The entire experiment package weighed around 2.2 pounds and was installed within the antenna cluster. As NASA speculated at the time:

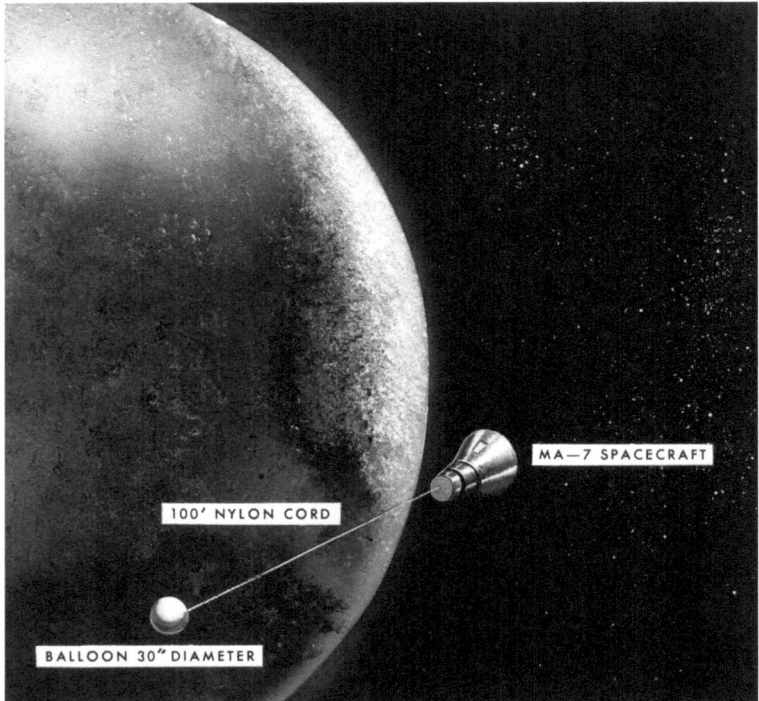

The tethered balloon experiment. (Photo: NASA)

"The operational plan calls for deployment of the balloon by the astronaut at the beginning of the second orbital pass. Output from the strain gauge and vocal response resulting from the visual observations will be recorded on tape. It is desired that the tethered phase last for nearly one orbital period; however, maneuverability of the spacecraft is necessarily restricted and earlier release may be required.

"The visual portion of this experiment will be concerned with the reflection characteristics of various colors in space, and the relative merit of these colors for optimum visibility. A correlation between observed and actual separation of the object from the spacecraft after release will be established. The aerodynamic portion will measure atmospheric drag and stability while deployed and a relationship between these parameters and object separation following release will be analyzed.

"The astronaut will observe the operation from the deployment sequence, through tethering, to release and separation, and any oscillations or gyrations will be noted. Photography of angular displacement, the various colors, and confetti dispersion will be provided for correlation with visual responses. The astronaut will orient the spacecraft in order to track the balloon's trajectory after it is released and photographs during this phase are requested when distances are recorded."[7]

Another experiment to be carried on the MA-7 mission, this time proposed by the Lewis Research Center, involved an examination of the behavior of liquids in the weightless environment. As NASA explained, various tests had been conducted using drop towers and aircraft in parabolic flight, but the test durations were inadequate for conclusive evidence. The orbital flights in the Mercury series offered the first opportunity to observe and photograph such behavior over an extended period. The forthcoming Gemini and Apollo programs would require a detailed analysis of weightless liquid in order to design fluid storage tanks, and the MA-7 experiment would help to establish the effects of surface tension, viscosity, mass, and liquid/gas volume ratios.

The apparatus to be loaded on *Aurora 7* had been qualified for flight at Cape Canaveral by the Lewis Research Center in Cleveland, Ohio, where it was designed and developed. It consisted of a spherical glass flask about three inches in diameter, with an internal one-inch standpipe that extended from the internal surface to slightly past center. The standpipe had three holes around its base to allow passage of the fluid. The flask was guarded on one hemisphere by a Lucite shield and on the other by an aluminum reflector. An O-ring was sandwiched between these two shields, so that in the event of a breakage of the flask, the liquid would not leak into the cabin. The glass flask had a volume of 300 milliliters and the liquid occupied 20 percent of this space, or 60 milliliters. The liquid consisted of distilled water, green dye, an aerosol solution to reduce surface tension, and a silicone additive to depress foam.

The MA-7 zero-gravity fluid experiment. (Photo: NASA)

The unit would be installed within the cabin of *Aurora 7* and observed by an astronaut-observer camera. The test device was located to the right and behind Carpenter's helmet, and he would periodically observe the status of the experiment using a hand-held mirror. One phase of the flight that was of particular interest to the scientists was the period during and immediately following retro-fire. It was theorized that in a zero-gravity condition, the liquid within the sphere would rise in the standpipe or capillary tube as a result of surface tension.[8]

STILL MORE EXPERIMENTS

Details of other proposed experiments were presented in the NASA Special Publication, *This New Ocean: A History of Project Mercury*. They were:

The Massachusetts Institute of Technology requested photographs of the daylight horizon through blue and red filters to define more precisely the Earth-horizon limb as seen from above the atmosphere. These findings would be particularly valuable for navigation studies in the Apollo program.

The Weather Bureau required information on the best wavelengths for meteorological satellite photography. John A. O'Keefe and Jocelyn Gill at the Goddard Space Flight Center and NASA Headquarters, respectively, wanted a distance measurement of the airglow layer above the horizon, its angular width, and a description of its characteristics. By way of explanation, airglow is an emission of light resulting from chemical reactions in the upper atmosphere. Various reactions produce light of different colors. In many cases, molecules of atmospheric gas are split by ultraviolet rays of sunshine. Then, when darkness comes, the gas molecules recombine, emitting light. The illumination of the sky at night usually comes from airglow instead of starlight.

A number of technical changes based on MA-6 mission results were made for MA-7, mostly involving deletions of certain equipment from the spacecraft in order to reduce precious weight. Kenny Kleinknecht's office eliminated the SOFAR [Sound Fixing And Ranging] capsule bombs and radar chaff recovery aids, which seemed unnecessary in view of the effectiveness that had been demonstrated by the SARAH [Search And Rescue And Homing] beacon and dye markers. Other items deleted included the knee and chest straps on the couch, which had bothered Grissom on his flight; the red filter in the window; the moderately heavy Earth-path indicator; and the instrument panel camera, which had already gathered sufficient data.[9]

While these experiments might not have overly interested the public at large, the one thing that truly provoked their curiosity and speculation was John Glenn's earlier reporting on the mysterious space "fireflies." Were they simply a phenomenon created by one of the many functions of the spacecraft and its pilot, or were they indeed some previously unknown form of life in space? While the latter scenario was highly improbable, having a conclusive answer to the question was obviously a juicier teaser for the public than any towed multi-colored balloon or other performance-related experiments. "Whether or not NASA liked it, the fireflies were now a celebrity space phenomenon with lives of their own," was the observation in *For Spacious Skies*. "No one knew for certain, and the burden of discovery was on the next pilot out there."[10]

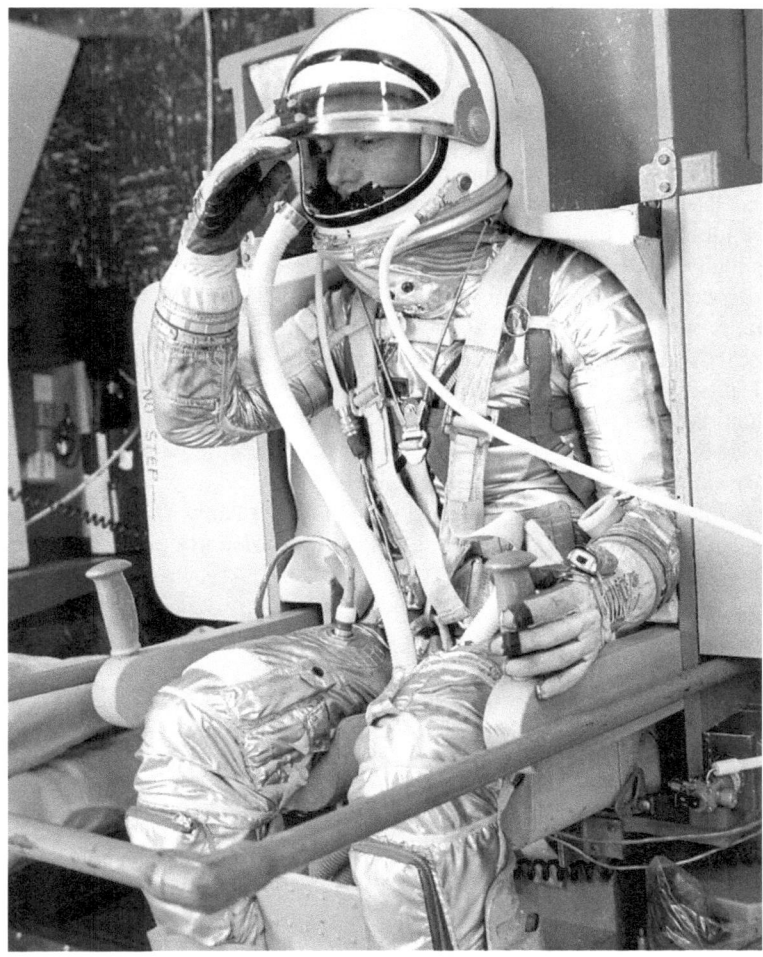

Scott Carpenter continues his training for the MA-7 mission. (Photo: NASA)

SUITING UP FOR SPACE

A definite essential and icon of America's space program was the silvery pressure suit worn on all six flights by the Mercury astronauts. In 1959, NASA had called for expressions of interest in the design and manufacture of individually customized pressure suits for future occupants of the cramped Mercury spacecraft. Among the companies competing for this contract were the David Clark Company of Worcester, Massachusetts; the International Latex Corporation (ILC) of Dover, Delaware; and the B. F. Goodrich Company, based in Akron, Ohio.

Aerospace manufacturing company B. F. Goodrich had a long and interesting background in designing wheels, brakes, and landing gear for airplanes, but they were flexible in what they could design and manufacture, and were readily capable of diversifying into other areas of aviation safety and pilot protection. In 1934, for example, the company was asked to design a pressurized rubber suit for pioneering aviator Wiley Post, who was keen to set new altitude records above 40,000 feet but needed a garment that would maintain pressure around his body by having air pumped into it. The assignment was handed to their 35-year-old mechanical engineer Russell S. Colley. Following several attempts, he finally came up with a suit made from rubberized balloon fabric and an aluminum helmet with a removable off-center faceplate specially tailored to the one-eyed Post's good eye. Colley's wife then stitched the $75 suit together on her sewing machine. After tests and modifications were completed, Post eventually reached 47,000 feet wearing the Goodrich pressure suit. Some years later Akron's *Beacon Journal* newspaper lauded Colley as "the first tailor of the space age."

In 1959, after an extensive evaluation program of three full pressure suits, the newly formed NASA space agency awarded the suit contract to B. F. Goodrich, agreeing to purchase 20 suits for the sum of $3,750 apiece, or about $29,000 in today's terms. The initial suit evaluation was conducted by the Aeronautical Systems Division of the Air Force's Aerospace Medical Laboratory. The first astronauts to be fitted with their individually tailored suits in October of that year were John Glenn and Wally Schirra. The pressure suits used in the Mercury flights were modified versions of the Navy's MK-IV full pressure suit, and various recommended changes were made as the Mercury program progressed from suborbital to orbital missions.

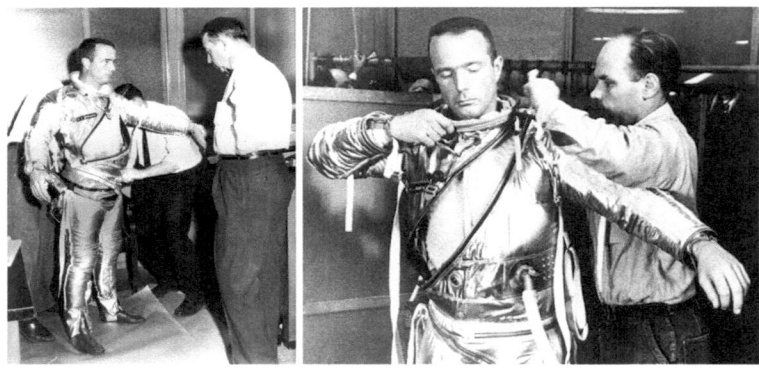

Scott Carpenter is fitted for his individual pressure suit by technicians at B. F. Goodrich. (Photos: NASA)

Each suit consisted of five basic components: the suit torso, gloves, boots, helmet, and undergarment. The suit torso was a closely fitted coverall custom-tailored for each of the astronauts, which covered the entire body apart from the head and hands. It had a two-ply construction comprising an inner gas-retention ply of neoprene and neoprene-coated nylon fabric, and an outer ply of heat-reflective, aluminized nylon fabric. The helmet was attached to the torso section by means of a rigid neck ring. Altogether, there were 1,600 custom-made parts in each suit. The suits were very snug on the astronauts, and the following year Colley revealed that this created a small dilemma. "We get the suit very carefully made – a perfect fit. And then the astronauts go on the banquet circuit and put on weight. It's a real problem."[11]

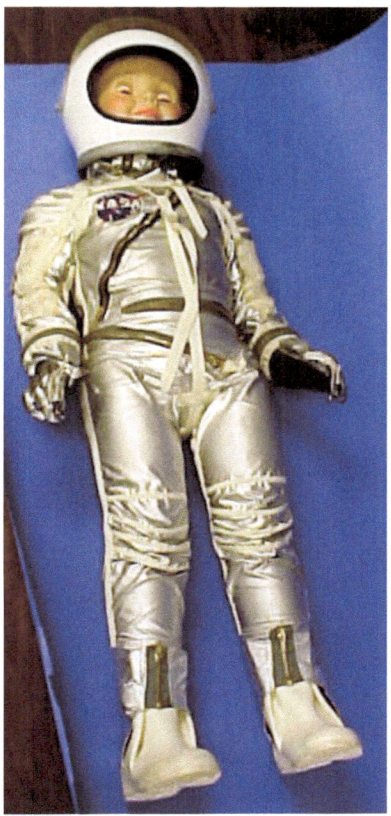

B. F. Goodrich produced a novelty item associated with Project Mercury. It was a doll dressed in a suit made from the same materials as the Mercury pressure suits. Only nine were made – one each for the seven astronauts, and one each for the chief suit engineer and product manager. (Photo: U.S. Air Force)

Navy Petty Officer and hospital corpsman Alan Michael Rochford, son of a colonel in the Air Force, was stationed with the Navy at Pensacola, Florida from 1958 to 1960, with much of his work centered around a low-pressure chamber. Among other disciplines, he was involved in teaching pilots the use of oxygen equipment, throwing different malfunction exercises at them, and conducting hypoxic demonstrations. During this experimental period the Mercury astronauts arrived in Pensacola, accompanied by retired hospital corpsman Glen Shumake. He was looking for corpsmen thinking of resigning from the Navy who might be interested in working in a low-pressure chamber established in Hangar S at Cape Canaveral. Although 23-year-old Rochford had already made tentative plans to leave the Navy, this sounded like an attractive proposition, so in August 1960, prior to his discharge, he applied for the job.

In late October a letter arrived from NASA asking Rochford if he was still interested in joining the space agency. He responded straight away, saying that he was, and a week later received a phone call asking if he could begin work at Langley Field, Virginia the following Monday. Thus began a 38-year career working for NASA, which encompassed every NASA human space program from Mercury on, through many of the Space Shuttle missions.

While working at Langley Field, Rochford was sent to the Cape with a team led by Glen Shumake to help set up and administer altitude chamber tests in Hangar S. The subjects of these tests were the Holloman Air Force Base-trained chimpanzees, two of whom would be launched on precursory Mercury missions ahead of the first human ballistic and orbital space flights. As he observed during a 1998 oral history interview, they left all the animal handling tasks to Air Force personnel from Holloman.

"The chimps had a little capsule that they were placed in, and we acted as chamber observers and chamber operators in Hangar S. So our job was to monitor. I remember there was a Beckman gas analyzer in the altitude chamber, in the airlock there, and our job was to monitor the gas analyzer. Glen Shumake ran the chamber …. So that's how I got into the NASA system."[12]

Meanwhile he had also been working at Langley's suit lab with pressure suit technician Joe Schmitt. While there, he became increasingly involved in working with Schmitt and the future astronauts.

"When we weren't in the suit shop … Glen Shumake, Tom Gallagher and myself would go down to the Cape on temporary duty to support some chamber runs, and then once we got through the chamber runs with the chimps, we went back to Langley. Then back down to the Cape with the astronauts. So I did a lot of traveling in those days. In fact, my whole career, I've traveled. But, I was a single guy, so it wasn't too bad."[13]

As Rochford revealed, each of the astronauts would fly up to the Goodrich plant in Akron, where he then, "Gets into his suit, pressurizes his suit, checks mobility with the suit inflated and soft suit, checks for pressure points, etc. If he is a happy camper he doffs his suit and flies home. Goodrich will then send it to us." And what if there were any further problems with the suit? "We could make adjustments to his arms, legs and back with lacing cord, but for major problems – such as a tear in the zipper – we would have to send it back to the factory."[14]

Rochford said each of the Mercury astronauts had three suits to work with. "They had two flight suits we call prime and a backup, as well as a training suit. The training suit was used primarily for dirty stuff like water egress training, that sort of thing. The flight suits were used for chamber runs and simulator training, and other activities. We would treat the backup suit just like a flight suit. In other words, if we made a modification to the flight suit, we'd do the same thing for the backup flight suit because the crew would be using both. They'd find that one fit just a little more comfortably than the other suit. As we got closer to flight time, the crew would choose, 'This is the suit that I want to fly.' So we would make the other one the backup flight suit. We tested those suits on an equal basis, because if something happened to the main suit on launch morning, we would have the other suit to fall back on."

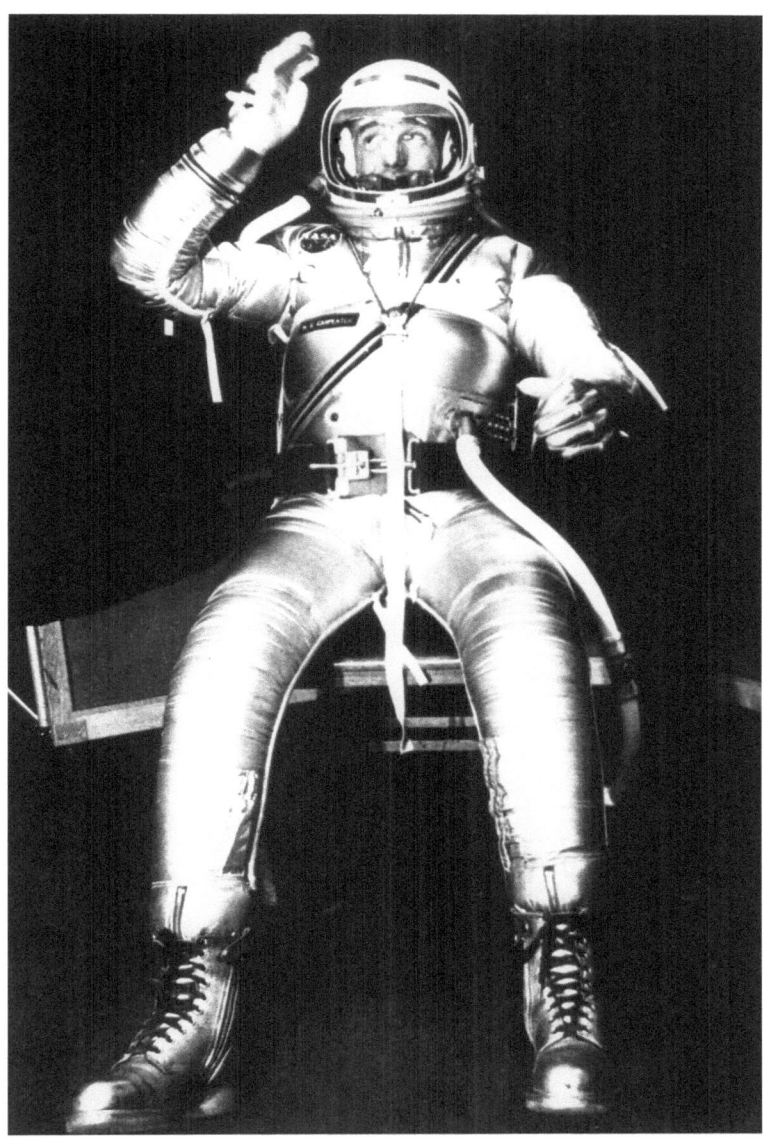

At the B. F. Goodrich factory, Scott Carpenter has his space suit pressure-tested. (Photo: NASA)

NASA suit technician Joe Schmitt in discussion with Scott Carpenter. (Photo: NASA)

Joe Schmitt was the prime suit technician for the Mercury flights, with Rochford acting as his backup. "Basically, I worked right along with Joe in supporting the crew in their training over at Mission Control at the Cape, and that's where the majority of the training took place, at the Cape. We supported all their suited activities. On launch morning, I was right there in the suit room with Joe while he suited up the crew."[15]

When asked if they stayed up the evening before a launch, Rochford said that was not the case. "We knew the time line, so we got enough sleep." Before suiting up the astronaut, they would conduct pressure tests on the empty suit as a final check before dressing the astronaut. The Mercury suit was fairly easy to get into, according to Rochford, and once they had the astronaut zipped up, they put the crewman's helmet and gloves on and plugged him into a machine that would circulate cooling air. He would climb into a pre-formed couch and they would run a manned leakage test on him. Once that was completed, he would go out fully suited, carrying a portable Sawyer battery-powered ventilator that was cooled with ice cubes.

Hangar S at Cape Canaveral. (Photo: NASA)

"We were on the second floor of Hangar S. We'd walk down the stairs … I would follow Joe, staying out of the camera range. [Laughter] Anyway, we'd go down into the transfer van with the medical doctor and one of the other astronauts. We'd go into a big semi-van that was set up with bioinstrumentation, a recliner chair, and a ventilator to keep the astronaut cool. As he sat in the recliner, we'd hook him up to ventilation, and he'd chit-chat generally among the doctor, Joe, or the other backup astronaut as we headed to the pad. Then Joe, the astronaut, and, I believe, the medical doctor would go up to the elevator to the spacecraft level. I would remain in the van in the event that they needed some backup equipment, which we carried in the van."[16]

For the MA-7 mission, Joe Schmitt had allowed Alan Rochford to take on a more senior role, but the flight delays meant he had to forgo the suiting-up honor due to an unavoidable conflict of dates.

Several days prior to his flight, Scott Carpenter is photographed suiting up in the Hangar S crew quarters, assisted by suit technician Alan Rochford. (Photos: NASA)

"I was supposed to be the prime technician for Scott Carpenter … that was supposed to have been my first flight … but what happened was John Glenn's earlier flight kept slipping and slipping and slipping, and Alice and I had set a wedding date for May 26, 1962. The flight kept slipping and finally … Scott was very understanding. He sent me a very nice letter after his mission (I still have that letter). I had talked to Joe, and I said, 'Look, I've had this all planned out, and I've got people coming from all parts of the country … do you mind if I complete these wedding plans?' Joe, gracious as he was, said, 'Sure. No problem.'" Two days after Scott Carpenter made his Mercury flight on May 24, Alan and Alice were married in Pennsylvania.

When asked whether he had any particularly fond memories to share, Rochford said, "I remember when Scott Carpenter was up at Johnsville, Pennsylvania. This was a little later in the Mercury program. He was there evaluating another suit of ours. This was in '63, because our son, Tom, was born on July 10, 1963. My wife, her parents and our son (who was around a month old) came over to the centrifuge and I introduced them to Scott. We plopped our son in his arms while he was fully suited and had a picture taken. Tom still has that picture. I thought that was pretty neat."[17]

On the following Mercury flight, MA-8, Alan Rochford became the prime suit technician for Wally Schirra, working hand-in-hand with Joe Schmitt, who had elected to be the backup technician for this flight. As Rochford recalled when asked if this was a one-off promotion to the prime level, and whether Schmitt would then have retained that position, "We would have rotated."[18]

In 1963, following the end of the Mercury program, NASA asked Rochford to transfer to Houston and continue working at the Manned Spacecraft Center. "Basically I closed up shop, transferred a lot of things from the Cape to Houston, and then I drove from Cape Canaveral

Pictured are, from left to right: Al Rochford, Scott Carpenter holding baby Tommy Rochford, Dominic Thomas, Alice Rochford, and Margaret Thomas. Dominic and Margaret Thomas were Alice's parents. (Photo courtesy Tom Rochford)

to Houston, and went to work at the Lane Wells Building in Houston because the site was still being built. That's where we started working on suits for the Gemini program."

In December 1997, following the launch of shuttle mission STS-87, Alan Rochford retired from NASA after 41 years of dedicated service to the space agency.

FACING THE MANY UNKNOWNS

Speaking at an American Institute of Aeronautics and Astronautics (AIAA) Conference in Dallas, Texas, on 22 April 1963, Scott Carpenter said that his training for MA-7 gave him the skill, confidence, and experience necessary to successfully complete his mission.

"In April of 1959, when the astronauts joined NASA, we were told that in our flights we would be subjected to a frightening combination of stresses, resulting from vibration, heat, cold, humidity, noise, acceleration, weightlessness, high concentration of carbon dioxide, immobility, disorientation, radiation and – hopefully – landing shock. Our eyes, inner ears, cardiovascular and respiratory systems, and even our very intellect were considered suspect by many."[19]

In recognizing that the astronauts would encounter many challenging obstacles in their flights, Carpenter explained how NASA put together a training program to familiarize them with these potential areas of concern. The program would prove to be of great benefit in providing accurate and representative experience. The most valuable training aid for the astronauts was the Procedures Trainer, which was a full-size replica of the Mercury capsule with controls and displays linked to an external computer. They also spent time building up their tolerance to g-forces on the Johnsville centrifuge, as well as undergoing desert, jungle, and water training, and experiencing brief periods of weightlessness during zero-g airplane flights. Instruction was given in SCUBA diving techniques (an acronym for self-contained underwater breathing apparatus), capsule egress training, and star recognition.

Carpenter in the Mercury spacecraft Procedures Trainer. (Photo: NASA)

"In retrospect," Carpenter added, "the only unknowns that existed in my mind after this very thorough training program was completed were: what would be the effect of prolonged weightlessness, and what reactions might be caused by various untried combinations of these stresses?"[20]

Carpenter's participation in the spacecraft pre-flight check-out activities enabled him to become fully familiar with *Aurora 7* and its launch vehicle systems. As Glenn's backup pilot he had spent 79 hours and 30 minutes training in the MA-6 capsule, and this, combined with 31 hours and 30 minutes in his MA-7 capsule, provided valuable training time. Additionally, he completed numerous exercises in the Procedures Trainer and the ALFA training device from 25 March to 22 May. During this intense period he spent a total of 70 hours and 40 minutes accomplishing 114 turnarounds, 92 retrofires, as well as experiencing 143 simulated systems failures of various types. The emphasis during these simulations was on practicing specific attitude maneuvers and rehearsing in-flight activities. Carpenter also received training on failure detection and correction, which usually resulted in an abort or early re-entry. He participated in several launch abort and network simulations during which the mission rules were discussed and rehearsed at length.

A vital system incorporated into the Mercury capsules was the Attitude Stabilization and Control System (ASCS) that fellow astronaut Gus Grissom had helped to perfect. Carpenter referred to the ASCS as "the brain of the capsule." It would, however, cause him many problems on his forthcoming mission.

"This device is crammed with sensitive gyros and horizon scanners which keep the capsule lined up at the proper angle with respect to the Earth, and it manipulates a total of 12 attitude-control nozzles which are part of the automatic system. Some of these nozzles give us 24 pounds of thrust to make major corrections or changes in the capsule's attitude, and some put out only six pounds of thrust to make the more minor corrections.

Aurora 7 is raised to the top of the launch gantry, where it is mated to the Atlas 107D booster. (Photos: NASA)

We can choose the thrusters we want to use if we take over manual control. The ASCS is capable of some uncanny feats. For example, using the horizon scanners, which are infrared eyes that determine the position of the horizon outside, the ASCS senses automatically how far the capsule may have moved from its programmed attitude in yaw, pitch or roll, then sends electronic impulses out to the solenoids to activate whatever fuel valves are needed to get the capsule back into proper alignment. The valves feed the nozzles which squirt out the decomposed fuel to make the actual corrections."[21]

Insertion techniques with the Mercury capsule on the launch gantry. (Photo: NASA)

NASA's Director of Operations Walt Williams (center) gives a pre-flight briefing with the MA-7 technical staff. John Glenn and Scott Carpenter are on the left. (Photo: NASA)

In preparation for the MA-7 mission, four of Carpenter's fellow Mercury astronauts were assigned to different CapCom stations around the world. Later on in the space program, communications with orbiting astronauts could be carried out from the Cape's Mission Control Center, but in those early days they had to be positioned below the orbital track. While Glenn and Schirra remained at the Cape in order to support Carpenter, Deke Slayton traveled to Muchea in Western Australia, Alan Shepard was assigned to Point Arguello in California to make contact with Carpenter when *Aurora 7* passed over the west coast of the United States. Gordon Cooper was sent to the tracking station at Guaymas, Mexico, and Gus Grissom was much closer to home as the CapCom astronaut at Cape Canaveral. His job was to talk directly with Carpenter as the Atlas rocket carried his colleague into space, and later as the spacecraft passed over the east coast and headed out across the Atlantic.

FLIGHT DELAYS

While Scott Carpenter was serving as backup pilot to John Glenn for the MA-6 mission, they had to endure a number of delays stretching over several months. As the time for his own flight drew near, Carpenter began to experience a similar set of technical delays, which he accepted with good grace because it gave him additional time to train for the experiment-packed mission.

On 7 May 1962, preparations for the flight (then some eight days away) were going well and preliminary checks, according to NASA officials, were "right on schedule." This was despite concerns raised when an Atlas F rocket exploded on 9 April due to the catastrophic failure of the liquid oxygen turbo pump just one second after lifting off from the Cape's Launch Complex 11. Following consultations with its engineers, the space agency decided not to modify the safety mechanism for ejecting the astronaut in an emergency during the first 150 seconds of flight, believing the system to be adequate. This decision eliminated a further delay of two weeks.

Late in the evening on Thursday, 18 May, with the rescheduled launch date just two days away, it was announced that the MA-7 mission would be delayed until the following Tuesday in order to improve the reliability of *Aurora 7*'s parachute system.

Launch simulations with the astronaut aboard took place atop the Atlas rocket. (Photo: NASA)

The announcement followed a mission review meeting attended by Scott Carpenter, Operations Director Walter Williams and other top Mercury officials. Their discussion had centered on a worrying event that occurred towards the end of John Glenn's flight, when the small stabilizing parachute on *Friendship 7* had been released and unfurled prematurely at 27,000 feet instead of the intended 21,000 feet. At the time, NASA officials had decided it was just a random failure and would not reoccur, so no changes were implemented in the same system on *Aurora 7*. But at Thursday's hastily convened meeting, Williams conducted a complete review of the MA-6 mission for any overlooked potential problems, and began to ask questions about the premature parachute deployment. He wanted to know if there was even the slightest possibility that the drogue parachute could pop out even sooner – perhaps even in Earth orbit. If this happened it would burn up during re-entry and effectively lose its value as a stabilizing force for the ocean-bound capsule. As the engineers were unable to furnish Williams with a satisfactory explanation after a 2.5-hour debate, he decided at 10:30 p.m. to postpone the flight until Tuesday the 22nd so that a backup parachute deployment switch could be inserted into the spacecraft.

The postponement to the following Tuesday actually gave some unfavorable Atlantic weather time to clear away. Weather experts had given the flight only a 50-50 chance of taking place on the Saturday due to winds and high waves in two ocean recovery zones.

Asked the next morning why nothing had been done about the parachute system in the weeks since Glenn's flights, NASA's public affairs spokesman Lt. John ("Shorty") Powers had replied, "In evaluating a flight, there are 90 million things to do. When it comes to a countdown, this is the time people toe the mark and ask why this or that didn't work before." He indicated that previously, officials were apparently satisfied that it was simply a random malfunction, perhaps generated by a rocking motion of the spacecraft during re-entry, and that subsequent tests had proven the system satisfactory.

The correction involved installing a second aneroid barometer switch which was sensitive to altitude changes. It meant that when atmospheric pressure reached a certain point – in this case 21,000 feet – the switch would expand and release the stabilizing drogue chute. The old and new switches would be hooked in series so that both functioned before the parachute unfolded. A similar switch would then release the main 63-foot parachute at 10,000 feet. Fortunately the main chute on Glenn's flight had worked perfectly, and no change was deemed necessary for that system.

Powers added that weather could force a further postponement beyond the projected Tuesday launch, but he said that the weathermen were more optimistic about chances for good weather on the Tuesday because of long-range forecasts.[22]

Then there was talk that nuclear tests in the Pacific might threaten to hamper the launch. *Aurora 7*'s orbital path would carry Carpenter directly over both Christmas and Johnston Islands, potentially passing through an area affected by blasts from three hydrogen bombs that would be detonated outside the atmosphere.

"I honestly don't know when a test is planned," Paul Haney, also from NASA's Public Affairs Office, admitted. But he explained that no attempt to send Carpenter into orbit would be made on the same day as an atomic detonation because of what "unknown factors might do to communications involving the orbital flight." He added that Project Mercury officials were in touch with the Department of Defense, and would be advised of any scheduled tests that might occur. The interim plan was for the U.S. Atomic Energy Commission to suspend nuclear blasts for several days before the launch of the MA-7

Aurora 7 sits atop the Atlas rocket during pre-launch checkouts as launch delays continue. (Photo: NASA)

mission, giving sufficient time for any high-altitude radioactivity to dissipate. But Haney said the nuclear tests would have to continue if any technical delays forced a prolonged postponement of the launch. Haney asserted that the astronaut would not be subjected to any dangerous amounts of radiation, emphasizing, "We certainly wouldn't let him fly through an atomic burst."[23]

Concerning this latest postponement, Carpenter remained philosophical, and was quoted as saying, "This is part of the continuing process of greater reliability – taking advantage of past flight experience – it gives me a chance to work with confidence."[24]

Today, very few people know that a terrible tragedy occurred in the lead-up to the MA-7 mission. On 17 May 1962 near Nairobi, Kenya, thirteen airmen supporting the Mercury flight were killed when their C-130 Hercules transport aircraft crashed into a mountain. The airmen were attached to the 322nd Air Division, 40th Troop Carrier and

317th Consolidated Aircraft Maintenance at Cape Canaveral Air Force Station. They were part of support teams sent to Africa for emergency recovery along the flight path of the Mercury orbital flights, and had been assigned to Scott Carpenter's *Aurora 7* mission. The day after the fatal crash, another support crew based at Evreux, France was assigned to a second maintenance aircraft to enable the orbital flight to go ahead as scheduled.[25]

REFERENCES

1. Ed Buckbee email correspondence with Colin Burgess, 26 December 2014
2. Lester A. Sobel (Ed.) *Space: From Sputnik to Gemini*, Facts on File, Inc., New York, NY, 1965
3. Scott Carpenter and Kris Stoever, *For Spacious Skies: The Uncommon Journey of a Mercury Astronaut*, Harcourt, Orlando, FL, 2002, pg. 246
4. NASA, "Technical Information Summary for Mercury-Atlas Mission 7 (MA-7, Spacecraft 18)," (Undated) Manned Spacecraft Center, Houston, TX
5. Jocelyn Gill interview; NASA, "Summary Minutes: Ad Hoc Committee on Scientific Tasks and Training for Man-in-Space (Meeting Nos. 1, 2, 3)," 16 and 26 March and 18 April 1962
6. NASA News Release 62–113, "MA-7 Press Kit," 13 May 1962
7. NASA *Space News Roundup*, "MA-7 Experiment to Determine how Colors Reflect in Space, Using Balloon and Confetti," Vol. 1, No. 15, 16 May 1962, pg. 3
8. NASA *Space News Roundup*, "Behavior of Liquids in Space to be Studied During MA-7," Vol. 1, No. 15, 16 May 1962, pg. 2
9. Loyd S. Swenson Jr., James M. Grimwood and Charles C. Alexander, *This New Ocean: A History of Project Mercury*. NASA Special Publication-4201 in the NASA History Series, 1989
10. Scott Carpenter and Kris Stoever, *For Spacious Skies: The Uncommon Journey of a Mercury Astronaut*, Harcourt, Orlando, FL, 2002, pg. 250
11. Mark J. Price, "First Astronaut's Spacesuits Were a Marvel in Their Day," *Astronomy & Space/Space Exploration*, 20 February 2012
12. Alan M. Rochford, interviewed by Summer Chick Bergen for the Johnson Space Center Oral History Project, Houston, Texas, 15 September 1998. Interview revised by Rochford for this book
13. *Ibid*
14. Alan Rochford email correspondence with Colin Burgess, 31 December 2014 – 3 February 2015
15. Alan M. Rochford, interviewed by Summer Chick Bergen for the Johnson Space Center Oral History Project, Houston, Texas, 15 September 1998. Interview revised by Rochford for this book
16. *Ibid*
17. *Ibid*
18. Alan Rochford email correspondence with Colin Burgess, 31 December 2014 – 3 February 2015
19. NASA *Space News Roundup*, "Carpenter Says Mercury Gave Skill, Confidence, Knowledge," Manned Spacecraft center, Houston, TX, issue Vol. 2, No. 14, 1 May, 1963, Pg. 3
20. *Ibid*
21. M. Scott Carpenter, L. Gordon Cooper, Jr., John H. Glenn, Jr., Virgil I. Grissom, Walter M. Schirra, Jr., Alan B. Shepard, Jr., Donald K. Slayton, *We Seven*, Simon and Schuster, Inc., New York, NY, 1962, pp. 159–161
22. *Ocala Star-Banner* newspaper (Florida), unaccredited article, "Carpenter's Orbital Flight Delayed Until Tuesday To Improve Parachute System," 18 May 1962, pg. 1
23. *Daily Mirror* (Sydney, Australia) newspaper article, "A-blasts may block space trip," issue 16 May 1962
24. *Ocala Star-Banner* newspaper (Florida), unaccredited article, "Carpenter's Orbital Flight Delayed Until Tuesday To Improve Parachute System," 18 May 1962, pg. 1
25. *Gettysburg Times* newspaper, article, "Report 17 Killed When U.S. Plane Crashes in Africa: Space Flight Link," issue 17 May 1962, pg. 1

4

Aurora 7 in orbit

"During the days immediately before the launch was first scheduled to go, I was tense and not at all at ease. I did not like the waiting, and I felt I needed more time to work on the flight plan and with the special equipment I would use."[1]

For Scott Carpenter, the anxiety he felt was entirely understandable. He had been thrust into the complex MA-7 mission at short notice, when so many unknowns still existed, and the overcrowded flight program was already proving burdensome and cause for some concern. While he knew in his test pilot's heart he was quite capable of carrying out a successful mission, he simply felt he was not fully prepared for the April launch date, and time was rapidly slipping away with the world's eyes squarely on him.

"I was not convinced in my own mind that I would perform well under the stress of the flight, and one night I had real difficulty getting to sleep. I was not afraid of dying itself, but I hated the thought of losing the life of the father of my four children, and I regretted the many experiences I would miss. There were only a few people I really wanted to be with, and away from work itself I remained either alone or in the company of John Glenn and a few other close friends."

Then, to his enormous relief, there was a double postponement to the mission, the first one related to naval exercises in the Atlantic and the second from poor weather conditions. He seized the opportunity this gave.

"The [first] scrub gave me a chance to practice more with the flight equipment and to study up on a few things I was worried about. I built up confidence and began thinking again the way I'd been thinking for three years – that on a successful flight I could make a valuable contribution and that it would be a great experience for me. I ate and slept well. As the new launch date approached I kept waiting for the tenseness to return, but it never did. I had reached the crest of the hill and became part of the machine."[2]

© Springer International Publishing Switzerland 2016
C. Burgess, *Aurora 7*, Springer Praxis Books, DOI 10.1007/978-3-319-20439-0_4

PREPARING FOR LAUNCH

As the revised launch date of 24 May neared, the recovery force took up positions for any contingency. If the launch was successful but only one orbit could be achieved, the pre-planned landing area was 500 miles east of Bermuda. If MA-7 ended at the completion of two orbits, then the landing area was located 500 miles south of Bermuda. The planned splashdown zone if all three orbits were completed was approximately 800 miles south-east of Cape Canaveral. Should the mission end prematurely at the end of the first or second orbits, the plan was to transport Carpenter to the Kindley Air Force Base Hospital in Bermuda for a 48-hour rest and debriefing. The best result would be the achievement of the full three orbits, which meant he would emulate the mission of John Glenn and be taken to Grand Turk Island for a post-flight medical examination and debriefing.

Meanwhile, Rene Carpenter was about to break with tradition by taking their children to the Cape to witness the launch. On all three previous Mercury missions, the astronauts' families and close friends had watched the launch and recovery operations live on their television sets at home, but Rene and Scott had earlier decided that their children should be present at the Cape when their father was launched into the skies. It was a truly covert operation, as the Carpenters and NASA did not want the general press to get wind that the family would be in Florida (although some did find out through leaked information). The day before the launch, Rene and the four children were secretly flown to the Cape, where a rental car was waiting at Patrick Air Force Base. Rene had disguised herself with a headscarf and sunglasses, and while the two boys sat in the automobile the girls were hidden beneath blankets in front of the back seat, creating the impression of a regular family of three out for a drive. The Cape was teeming with carloads of tourists eager to watch the historic launch, so one more did not attract any attention.

Their destination was a secluded beach house located about ten miles south of Launch Complex 14, which offered the privacy the family desired. They had stayed at this house once before, in December 1961, and if they stepped outside they could look north to the launch pad from where Carpenter would be fired into space. One of two houses leased by *Life* magazine, it served primarily as a retreat for the astronauts. It was where Scott Carpenter had gone to relax and wait for news following John Glenn's launch. Under the existing financial arrangement with *Life* magazine, a photographer and journalist were given exclusive access to the family inside the beach house, specifically to record their reactions during the launch. Rene and the children would watch a television set tuned to the launch preparations, but when ignition was announced she planned to have them all run outside to watch the rocket's ascent.

The evening prior to his scheduled launch, Carpenter studied flight data for a while before retiring to bed just after 10 p.m. He was in a relaxed state, he opted not to take a sedative.

THE DAWN OF AURORA

On the morning of the flight, Carpenter was awakened by Dr. Howard Minners at 1:15 a.m., some 65 minutes earlier than Glenn had been for his MA-6 flight. He had managed around three solid hours of sleep. "There wasn't, at that time, a major effort to change your

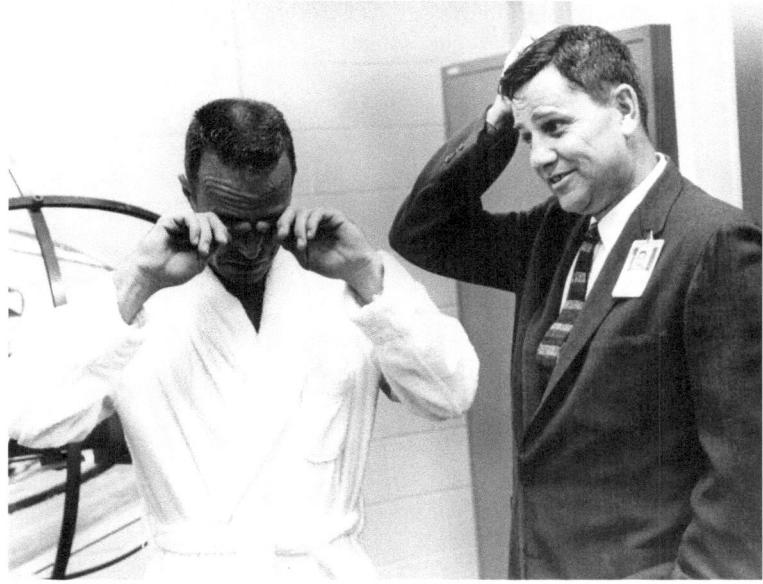

Rubbing the sleep from his eyes before breakfast on launch day, Carpenter is joined by NASA's Chief of the Pre-flight Operations Division, G. Merritt Preston. (Photo: NASA)

diurnal cycle, because the impact on it really wasn't significant. If it is, it's easily overwhelmed by the anticipation of future events. The adrenalin helps."[3]

After a shower and shave he slipped into a robe and sat down with Drs. Minners and Douglas and John Glenn for his low-residue breakfast, which consisted of filet mignon, poached eggs, strained orange juice and toast. Unlike John Glenn, he drank regular coffee at the breakfast rather than Postum, a rather bland coffee substitute from which all the caffeine had been extracted.

At 2:05 a.m., he was given a pre-flight physical examination by Bill Douglas, Howard Minners, and other specialists in internal medicine and neuropsychiatry, who had previously carried out extensive medical checks on him. His physical and mental status was deemed to be quite normal under the circumstances, and Dr. Minners reported that Carpenter was "in excellent physical condition, and his attitude is perfect." Biosensor placement on his body took place just after 2:40 a.m. The biosensor system consisted of two sets of electrocardiographic leads – ECG 1 (axillary, or armpit) and ECG 2 (sternal); a rectal temperature thermistor; a respiration-rate thermistor; and the blood-pressure measuring system.

While he was undergoing these pre-flight preparations, Carpenter received a short visit from fellow astronaut Gus Grissom, who had called into Hangar S to wish his colleague good luck before making his way out to the Capsule Communications desk in the Mercury Control Center.

At 2:55 a.m., word arrived from Operations Director Walt Williams at Mercury Control that everything was ready and the flight was on.

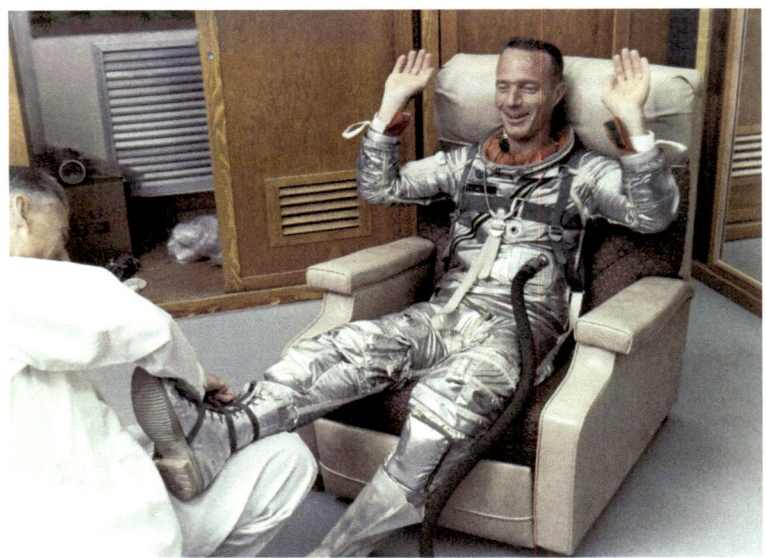

A relaxed Scott Carpenter shares a laugh with suit technician Joe Schmitt as his boots are laced up. (Photo: NASA)

In the next stage of pre-flight preparations, Carpenter was assisted into his 20-pound silver pressure suit by NASA suit technician Joe Schmitt. At 3:25 a.m. Schmitt conducted a final pressurization check on the spacesuit, and was pleased to note that the suit pressure was maintained following inflation, indicating there were no leaks.

At 3:40 a.m. NASA public affairs officer Col. John ("Shorty") Powers emerged from Hangar S to announce, "Scottie looks very, very good this morning." Two minutes later, carrying his portable Sawyer air-conditioning unit, Carpenter made his way downstairs and exited the hangar, accompanied by Schmitt and Dr. Minners. They made their way past photographers and well-wishers and climbed into the transfer van that would carry them the four miles to the launch pad. At this time, the launch was set for a 7:00 a.m. liftoff.

Right at 4:11 a.m. the van pulled up to the launch pad, which was ablaze with light from massive searchlights. Once there, Carpenter waited 19 minutes, ready to hear confirmation that it was time to leave the van and ascend the gantry. During this time he was visited by physiologist Dr. Rita Rapp from NASA's Life Systems Division, who talked briefly with the astronaut, but noticed no undue anxiety in him about what lay ahead.

Finally the word came, and at 4:38 a.m. Carpenter stepped out of the van to a smattering of applause and shouts of encouragement from the engineers and technicians gathered at the foot of the fuming Atlas rocket. After making his way to the gantry elevator, he stopped for a few moments to talk with the Launch Complex 14 project manager for

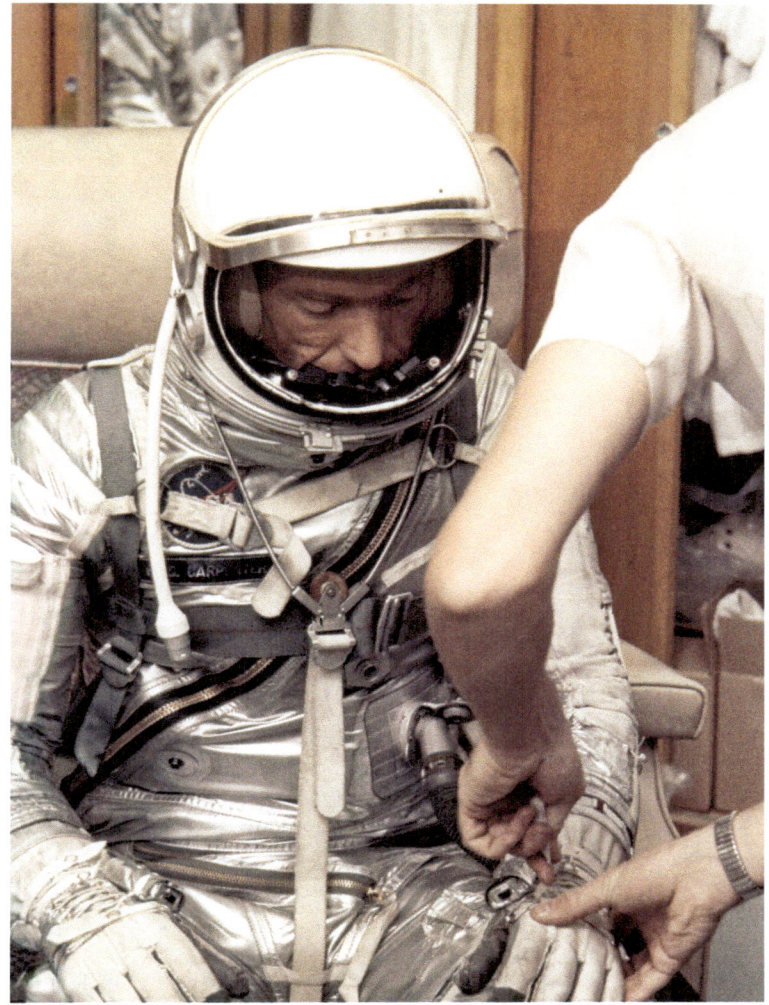

With his helmet now in place, Carpenter watches as Schmitt secures his gloves. (Photo: NASA)

General Dynamics, Byron B. G. McNabb, who said, "Welcome aboard," and wished him good luck. With that, the astronaut turned and entered the elevator for the 95-foot ride to the top of the gantry.

"It was dark when I rode the elevator up to the gantry to be inserted into the capsule. Wally [Schirra] was waiting to greet me after having checked out the capsule through the preliminary stage of the count. 'It's ready, Scott,' he said. 'It's all yours.'"[4]

McDonnell's pad leader Guenter Wendt also assured Carpenter that the spacecraft was ready to go, so he slipped off his protective overshoes and prepared to begin the adventure of his life.

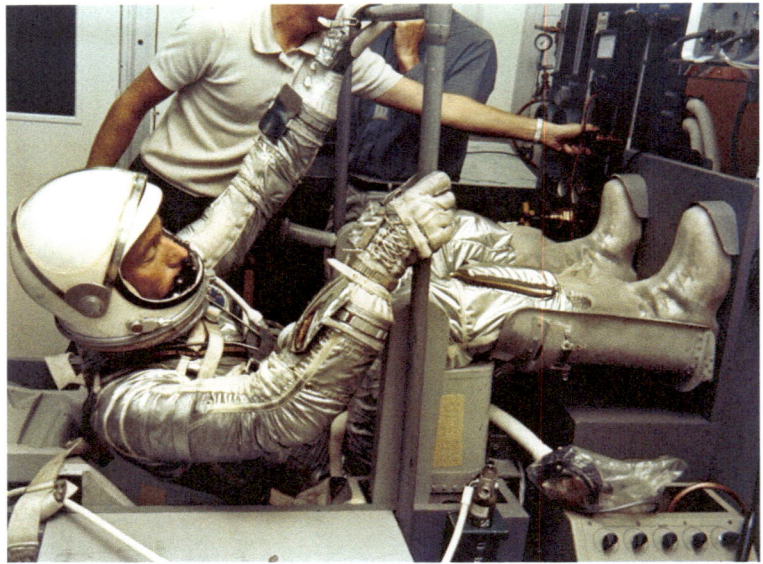

A final check of the pressure suit to ensure its integrity prior to leaving for the launch pad. (Photo: NASA)

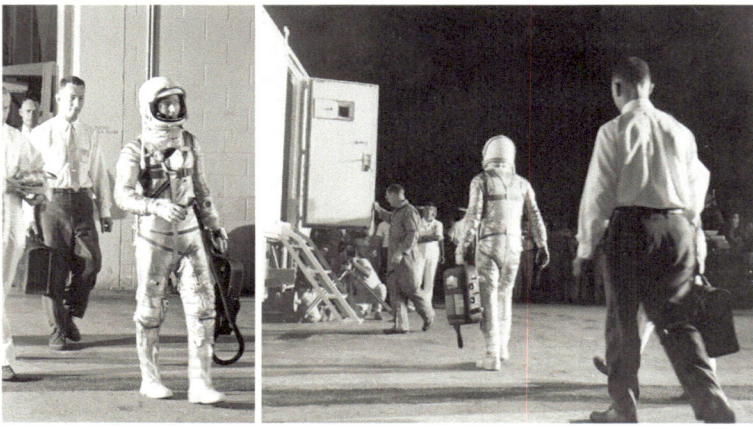

Ready to go. Carpenter makes his way out and into the transfer van, followed by Joe Schmitt and Howard Minners. (Photos: NASA)

Shrouding the Mercury-Atlas rocket, the red launch gantry is bathed in the brilliant light of a battery of searchlights. (Photo: NASA)

Carpenter's feet-first insertion into the claustrophobic confines of the spacecraft began at 4:44 a.m., as he carefully squeezed past all the gear mounted inside. One of the astronauts had once half-jokingly stated that you didn't climb into the Mercury capsule – you put it on.

Having practiced this insertion routine many times, Carpenter was soon comfortably positioned in his contour couch in a semi-supine position. Suit technician Joe Schmitt then took over, to run through the process of hooking up the ventilation hoses, instrumentation

On board the transfer van, Carpenter appears relaxed about the flight. (Photo: NASA)

and communications cables, biosensor leads and helmet visor seal hose, ending with securing the astronaut's shoulder and lap harnesses. Next, Dr. Bill Douglas completed his final inspection of the interior of the spacecraft and checked that Carpenter was satisfied, set and ready. Once everyone had finished, Wendt supervised the closure of the hatch and the torquing down of the 70 hatch bolts.

Carpenter steps down from the transfer van to begin his journey into space. (Photo: NASA)

Carpenter peeps inside Aurora 7 before the insertion process begins. (Photo: NASA)

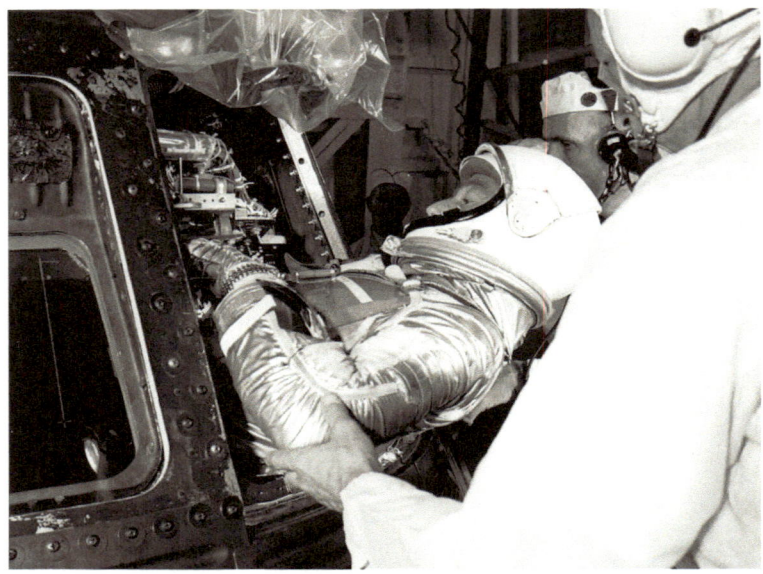

Carpenter is carefully inserted through the side hatch and into the cabin of *Aurora 7*. (Photos: NASA)

FINAL COUNTDOWN

On that glorious May morning everything looked good. Carpenter was in good health and ready to go. Atlas booster 107D and *Aurora 7* had both checked out perfectly, and there was a good forecast for the weather at launch time. During this period, the final pre-flight checks were completed, and as he waited, Carpenter performed frequent deep-breathing and muscle-tensing exercises.

"Through the small window above me, I watched the sky lighten. At T-6 minutes, visibility was still impaired by the usual morning haze, now thickened by smoke from distant swamp fires. We knew the Sun would soon burn away the haze, and Walter Williams, Project Mercury's operations director, decided that with so many things going right it would be best to wait a while."[5] As the countdown had proceeded without a hitch, Williams ordered a 15-minute hold to give the Sun a chance to burn off the lingering morning fog surrounding the launch pad and allow the pall of brush fire smoke from the Everglades to dissipate. The hold was extended to 30 minutes, and then to 45 minutes.

The delay allowed Carpenter to do something that he had earlier decided against. Shortly after seven o'clock, Rene Carpenter received an unexpected call in the beach house.

"If we had not had a 45-minute hold for weather the morning of the launch, I would not have called Rene from the capsule," explained Carpenter. "The countdown had been moving along perfectly, and I had too much to do to take time out for a call. Rene and I had agreed that such a last-minute conversation might be upsetting for both of us. But during the brief delay I decided that I wanted to call and that I could handle it."[6]

As Rene recalled, "Wally Schirra's voice came on the line first, and it had that peculiar sing-song quality that is characteristic of the tense conversations during countdown. I talked to Scott and then the kids chattered away to him … I was surprised at the call, for after Scott monitored John's call to Annie from the capsule, he was convinced it would be better for him if he didn't phone me."[7]

After an initial reluctance, Scott was glad that he had decided to call his wife, although he later confessed, "It got through to me only once, briefly, and tears came to my eyes. But it passed."[8]

At T-10 minutes the countdown resumed and as before proceeded smoothly. Outside the capsule the visibility was rapidly improving and the low, thin cloud cover was lifting.

"At work again I found myself amazed at my own calm," Carpenter related in a post-flight *Life* magazine article. "I felt a certain detachment, as if I could stand a little to one side and watch myself get ready. Perhaps this detachment is a defense against fear, much the way shock is a defense against pain, but I approached the moment the flight would begin with compelling curiosity. I remember from childhood that when my grandfather was dying of a stroke he said to his doctor, 'At last I'll know the great secret.' High on top of the Atlas, I was confident that everything was going to be all right, but I felt that I, too, was going to be let in on a great secret, that this experience I had looked forward to for so long would soon be here."[9]

As the countdown proceeded, Carpenter sat quietly, continually gazing around the panels of instruments, checking that everything was as it should be. There was time for some quiet contemplation, but not prayer. He believed he was as prepared as he could be for the task ahead, and so were the booster rocket and *Aurora 7*.

"During the pre-launch period I had no problems," he later stated. "The couch was comfortable, and I had no pressure points. The length of the pre-launch period was not a problem. I believe I could have gone at least twice as long. Throughout this period, the launch vehicle was much more dormant than I had expected it to be. I did not hear the clatter that John Glenn had reported."[10]

Downrange and across the world, recovery forces were in position and alerted to the fact that launch time was close. In all, 20 ships, 110 aircraft, 13,000 Army, Navy, Air Force and Marine Corps personnel were prepared for their role in the historic event.

Atlas 107D is poised for liftoff as the count winds down. (Photo: NASA)

Meanwhile, listening in to the announcements from the Mercury Control Center, excited onlookers at the Cape heard the voice of Col. Powers as he ran through the final few seconds of the countdown. "T-19 seconds and holding momentarily … T-15 and counting … 10, 9, 8, 7, 6, 5, 4, 3 …" Having gotten slightly ahead of the clock, he repeated one number, "3, 2, 1, ignition; liftoff!"

"THE CLOCK HAS STARTED"

It was 7:45 a.m. when two small stabilizing rockets at the foot of the frost-covered Atlas booster shrieked into life, followed by the ignition of the three main engines. Carpenter immediately started to mentally store his impressions, so that he could later describe the associated sensations. "When the ignition signal was given, everything became quiet. I had expected to feel the launch vehicle shake, some machinery start, the vernier engines light off, or to hear the lox [liquid oxygen] valve make some noise, but I did not. Nothing happened until main engine ignition; then I began to feel the vibration. There was a little bit of shaking. Liftoff was unmistakable."[11]

To onlookers crowding the nearby beaches and roads, and millions watching on television sets across the nation, it all seemed to happen very slowly at first as the mighty Atlas gathered up power. A massive grey-white cloud belched outwards from the base of the rocket, quickly billowing higher than the pad-bound missile. Then explosive steel clamps fell away, and the bird was free. "I feel the liftoff," Carpenter reported. "The clock has started."

CapCom Gus Grissom, with John Glenn standing behind him in the nearby blockhouse, acknowledged the transmission with a relieved, "Roger."

Once the Atlas had been unleashed from its shackles it began to rise, ponderously at first and then ever faster above a golden plume of liquid flame, powered by 360,000 pounds of thrust. The rocket's ultra-thin skin was frosted with condensation from the super-cold liquid oxygen propellant within, and as it accelerated through the morning air and remaining haze, the icy condensation was forced to stream downwards. On a nearby and secluded stretch of beach, highly respected *Life* photographer Ralph Morse snapped a dramatic series of images of Rene Carpenter and her children standing outside the beach house, watching as the Atlas ripped a fiery path through the Florida sky. One of his photographs of Rene would grace the cover of the magazine the following week.

The Atlas quickly cleared the launch tower. While it was not as noticeable on black-and-white television sets, those watching the launch at the Cape were enthralled by the dazzling conflagration of fire and smoke trailing below the Atlas. Some even pointed out the two small vernier engines gushing flame like twin blowtorches at the foot of the ascending rocket. Then a vast crackling sound swept across the Cape, and everyone watched in awe as Scott Carpenter climbed ever higher into the blue sky and into the history books, leaving behind a blackened, smoldering launch pad.

"The launch was a snap," Carpenter later recorded. "We rose with very little vibration in the capsule. In fact, the whole period before I was placed in orbit was gentler than I'd anticipated. The engines made a big racket, but there was no violent trembling of the

Amid billowing clouds of smoke the Atlas rocket roars into the morning sky. (Photo: NASA)

whole structure …. As we climbed I did notice a distinct swaying motion of the whole machine. John Glenn had reported this on his flight and had said it felt to him as if he were on the end of a springboard. It did not feel that way to me; it seemed rather that we swayed off to one side and stopped abruptly, then swayed back to the other side and stopped again. But the motion was not alarming."[12]

Crowds on nearby beaches watch in wonder as *Aurora 7* takes to the skies. (Photo: NASA)

Elsewhere in the United States, commuters come to a standstill in Grand Central Station to watch the MA-7 launch on a huge television screen. (Photo: NASA)

Broadcasting from the Cape's Mercury Control Center, John Powers told the world that the "MA-7 trajectory is okay. We are now one minute and seventeen seconds into the flight. Scott Carpenter reports his fuel and oxygen quantity steady. The cabin pressure decreasing on schedule."

As Powers was offering his commentary, Carpenter observed the light blue sky begin to darken around him as the Atlas continued to accelerate. He was now being forced into his formfitting couch by a front-on force eight times that of gravity. "I remember thinking to myself as I watched the altimeter needle wind up to 70,000 feet, then 80,000, then 90,000, 'What an odd place to be, and going straight up' – and I had about 560,000 feet still to go."[13]

In the Mercury Control Center, CapCom Gus Grissom monitors the progress of Carpenter's flight. (Photo: NASA)

EARLY PROBLEMS

Although the launch was later regarded by many as near-perfect, trouble was already brewing for the astronaut, as later recorded in his autobiographical *For Spacious Skies*.

> No one noticed at the time – there was no dial to measure its functioning – but the capsule's pitch horizon scanner had already started malfunctioning. The Mercury capsule was chock-full of automatic navigational instruments, among them the PHS [pitch horizon scanner], which does just what the name implies. It scans the horizon for the purposes of maintaining, automatically, the pitch attitude of the capsule. For MA-7, however, the PHS would feed erroneous data into the Automatic Stabilization and Control System (ASCS), or autopilot. When this erroneous data was fed into the ASCS, the autopilot responded, as designed, to fire the pitch thruster to correct the perceived error. This in turn caused the spacecraft to spew fuel from the automatic tanks.[14]

Within 90 seconds of liftoff, *Aurora 7* had climbed through the majority of the Earth's atmosphere and was rapidly passing into the deep blackness of space. Then, just over two minutes into the flight and right on schedule, came the booster engines cut-off (BECO).

"Three seconds later, staging occurred. There was no mistaking staging. Two very definite noise cues could be heard: one was the decrease in noise level that accompanied the drop in acceleration; the other was associated with staging. At staging there was a change in the light outside the window and I saw a wisp of smoke."[15]

Staging was a crucial phase of the flight. As the Atlas rose into the sky, its fuel was rapidly depleted. Once the rocket became light enough, the two main engines, one either side of the central sustainer engine, having done their heavy-lifting job, were shut down and then jettisoned in the phase known as staging. The sustainer would continue to propel the capsule towards orbit with the two small vernier engines doing the steering.

Another 20 seconds on, the escape tower, which was no longer needed, was jettisoned. Carpenter later stated that he felt a bigger jolt than at staging, and the tower was gone in a second, streaking away "like a scalded cat." As he watched the tower receding toward the horizon and slowly begin to fall back to Earth, he reported that his status was good.

The next major event, right on time five minutes into the flight, was the sustainer engine cut-off (SECO). This was rapidly followed by the next event, in which explosive bolts were detonated to sever the clamps and metal connections that held the *Aurora 7* spacecraft to the now fuel-gutted booster. Then three solid-rocket posigrade thrusters on the blunt end of the capsule fired briefly to propel it safely away from the Atlas at 24 feet per second. Carpenter confirmed the posigrade thrust.

"I am weightless and starting the fly-by-wire turnaround," he reported. As the periscope came out, he received word from Grissom that everything was "go" and the spacecraft had a provisional seven-orbit capacity. "Sweet words," he acknowledged. It was a milestone to celebrate in this early phase of the flight; he wasn't going to have to head home early.

As he recalled in the astronaut book, *We Seven*, "The first thing that impressed me when I got into orbit was the absolute silence. One reason for this, I suppose, was that the noisy booster had just separated and fallen away, leaving me suddenly on my own. But it was also a result, I think, of the sensation of floating that I experienced as soon as I became weightless.

All of a sudden, I could feel no pressure of my body against the couch. And the pressure suit, which is very constricting and uncomfortable on the ground, became entirely comfortable. The pressures were all equal; even a change of position made no difference. It was part of the routine to report this moment to the ground as soon as it came. It was such an exhilarating feeling, however, that my report was a spontaneous and joyful exclamation: 'I am weightless!' Now the supreme experience of my life had really begun."[16]

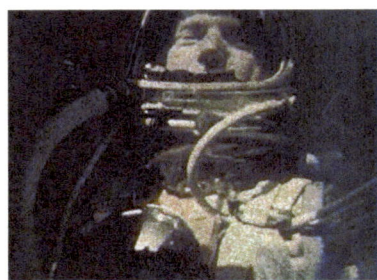

Scott Carpenter begins his three-orbit journey around the world. Unfortunately film footage of the astronaut was badly damaged by sea water after splashdown. (Photo: NASA)

Aurora 7's orbit, inclined at an angle of 32.5 degrees to the equator, had an apogee (or high point) of 167.4 miles and a perigee (low point) of 99.2 miles. The spacecraft would circle the Earth once every 88.3 minutes.

Carpenter radioed back that the capsule turnaround had been completed. The blunt end of the capsule behind him was now pointing along the track he would follow. He had used the manual control system to execute this maneuver, and was delighted to note that *Aurora 7* was responding beautifully. When he moved the control stick, small hydrogen peroxide thrusters on the outside pushed the capsule into whatever attitude he wanted. "I'm pitching down," he reported. "I have the Moon in the center of the window, and the booster off to the right slightly."

It was time to get down to work. Prior to the flight, he had prepared a series of index cards on which he had written down his scheduled activities minute-by-minute for each orbit. Every card had a small piece of Velcro on the back so he could attach it to a corresponding surface in the cockpit. He then opened up his ditty bag, which contained experiment-related equipment that he would use during the flight, as well as bags of food items. Included in the equipment were his modified camera, the airglow filter for measuring the frequency of light emitted by the airglow layer, star navigation cards, and weather charts.

"First out was the camera, for I needed to catch the sunlight on the slowly tumbling booster still following the capsule. The camera had a large patch of Velcro on its side. I could slap it on the capsule wall when it wasn't in use. Velcro was the great zero-gravity tamer. Without it, the equipment would have been a welter of tether lines – my idea, incidentally, and not a very good one, for John's flight. He had ended up in a virtual spaghetti bowl full of tether lines and equipment floating through his small cabin."[17]

After further use of the maneuvering system he advised that the control system on fly-by-wire was "very good." He then added, "I have the booster in the center of the window now, tumbling very slowly."

The blurred object in this photograph is the spent Atlas booster falling behind *Aurora 7*, from which it had separated shortly before. (Photo: NASA)

In the early stages of the first orbit, with so much to do, he had not taken the time to look around, but when he did, the sight was overwhelming.

"There were cloud formations that any painter could be proud of – little rosettes or clustered circles of fair-weather cumulus down below. I could also see the sea down below and the black sky above me. I could look off for perhaps a thousand miles in any direction, and everywhere I looked the window and the periscope were constantly filled with beauty. I found it difficult to tear my eyes away and go on to something else. Everything is new and so awe-inspiring that it is difficult to concentrate for very long on any one thing .... Using the special camera I carried, I took pictures as fast as I could, and as I raced towards night at 17,500 miles an hour I saw the beginnings of the most fantastically beautiful view I have ever had – my first sunset in space."[18]

In his post-flight pilot's report, Carpenter would point out, "Unlike those on Earth, the sunrises and sunsets in orbit were all the same. The sharply defined bands of color at the horizon were brilliant. On the dark side of the Earth, I saw the same bright band of light just above the horizon which John Glenn reported. I measured the width of this band in a number of ways, and I also observed it through a special 'airglow' filter."[19]

THE WORLD THROUGH A WINDOW

Despite his heavy workload, as Carpenter came within range of the Canary Islands tracking station and the African continent came into view beyond, he took time to marvel at the many splendors outside his window and through the periscope.

"The brilliance of the horizon to the west made the stars too dim to see in the black sky. But I could see the Moon and, below me, beautiful weather patterns. But something was wrong. The spacecraft had a scribe line etched on the window, showing where the horizon should be in retro-attitude. But it was now above the actual horizon. I checked my gyros and told Canary CapCom my pitch attitude was faulty. Then I added an explanation – it was 'probably because of that gyro-free period' – and dismissed it. There were too many other things to do."[20]

On later reflection, Carpenter said that a thorough ASCS check early in the flight could have identified the fault. "Ground control could have insisted on it, when the first anomalous readings were reported. Such a check would have required anywhere from two to six minutes of intense and continuous attention on the part of the pilot. A simple enough matter, but a prodigious block of time in a science flight – and in fact the very reason ASCS checks weren't included in the flight plan. On the contrary, large spacecraft maneuvers, accomplished off ASCS, were specified, in addition to how many minutes the MA-7 pilot would spend in each of the three control modes – fly-by-wire, manual proportional, and ASCS."[21]

NASA Flight Director Gene Kranz was monitoring the flight in the Mercury Control Center. In his autobiography, *Failure Is Not An Option*, he explained the function (and failure) of the horizon sensors on the MA-7 mission.

"Horizon sensors detect the difference in infrared radiation between space and Earth. The sensors provide signals to update the gyros that control the pitch and roll, and *Aurora 7*'s pitch sensor was varying by as much as plus 50 to minus 20 degrees. Carpenter was the only one who could put all of the pieces together by comparing his instrument readings with the spacecraft periscope and the view of the horizon from the capsule window. If the readings were off, like a pilot in an aircraft, he could realign the gyros to the correct position."[22]

Passing over Africa, Carpenter photographed the cloud-covered west coast and the blue expanse of the Atlantic Ocean. (Photo: NASA)

"MA-7 was no picnic," Carpenter later observed in *For Spacious Skies*. "I had trained a long time, first as John's backup, and then for my own surprise assignment to the follow-on flight. To the extent that training creates certain comfort levels with high-performance duties like space flight, then, yes, I was prepared for, and at times may even have enjoyed, some of my duties aboard *Aurora 7*. But I was deadly earnest about the success of the mission, intent on observing as much as humanly possible, and committed to conducting all the experiments entrusted to me. I made strenuous efforts to adhere to a very crowded flight plan."[23]

As Carpenter was all too aware, his flight aboard the *Aurora 7* spacecraft was different in many ways to that of John Glenn. "His was a real pioneer mission," he later observed, "and he had to concentrate on proving the reliability of the machine. Because he showed that a man can handle the machine under very difficult circumstances, I was to have more freedom to measure, study and observe events which were taking place *outside* the capsule. I had many sciences to serve."[24]

While passing over the African continent he reported seeing and recognizing several landmarks, including vast rain forests, Lake Chad, and the island of Madagascar. He did express surprise at how much of the planet was covered by clouds. Having given his spacecraft and physiological report to the Canary tracking station, Carpenter found himself continually mesmerized by the glorious view outside.

From *Aurora 7*, Carpenter photographed the spectacular beauty of a sunset as seen from space. (Photo: NASA)

"Through the window, I could see the Sun actually dropping towards the western horizon. Right on the horizon as the Sun fell, a band of color stretched away for hundreds of miles to the north and south. It was a glittering, iridescent arc composed of strips of colors ranging from yellow-gold to reddish-brown, to green, to blue and then to a magnificent purplish blue before it finally blended with the black of the sky. The colors were all sharply separated and glowed vigorously, alive with light, and I watched the band narrow until nothing was left but a rim of marvelous blue. It occurred to me that I now knew what a new

Earth must look like from the Moon (with the Earth in line with the Sun and lighted from behind) – like a bright blue ring in the sky. I looked in the periscope soon after that and saw nothing but blackness ahead of me. I was on the dark side of the Earth."[25]

The moment the Sun disappeared, Carpenter took note of the time. It was 47 minutes and 34 seconds into his flight. Just over half a minute later he did a fuel check and was surprised at what he saw. "Oh dear, I've used too much fuel," he reported to the ground. He had been constantly maneuvering *Aurora 7* around in order not to miss anything, and placing the spacecraft in the best possible attitude for taking photographs.

While on the dark side of the Earth he observed the same bright band of light known as the airglow layer just above the horizon that had been reported by John Glenn on his MA-6 mission. In one of the most important experiments listed for his flight, Carpenter measured the width of this band in a number of ways, and also observed it through his special air-glow filter. In this way he was able to measure the luminous layer's distance from the horizon as well as the light intensity of stars seen through the layer.

In looking out through his window, he later reported that any future spacecraft should have their interiors designed in a different manner so that pilots could get a clear look at the starry sky. His difficulty stemmed from the cabin's panel lights, and particularly the light leak around the time correlation clock. With the tremendous array of instruments that he had to continually monitor, he could not turn the lights off or dim them. As a consequence, his eyes did not become adapted to darkness and he saw no more stars than were visible from the ground. Nevertheless, he could readily see and identify the major constellations and use them for heading information.

THINGS WARM UP

It became apparent around this time to Carpenter and those monitoring the spacecraft's systems that there was also a malfunction in the astronaut's suit cooling system, and the increase in heat was making him slightly uncomfortable. He reported the problem to Deke Slayton as *Aurora 7* passed over the Muchea tracking station in Western Australia, and Slayton suggested a different setting, which he tried. He increased the water flow into his suit and the temperature slowly decreased. It would rise again later in the flight, with Carpenter experiencing great difficulty in achieving the proper water flow setting for the suit's cooling system. Other than that problem, he was obviously in good condition and performing satisfactorily at this time in the flight.

As he continued to pass over the Australian continent, he was on observation alert for the planned sighting of four star-shell flares of one million candlepower that were being fired up from the Great Victorian Desert near Woomera. Ignited at 60-second intervals, each flare burned for 90 seconds. The intention was to report on how clearly something like that could be seen from space, but the experiment was thwarted by heavy clouds covering the firing area, so he saw nothing.

"But I maneuvered the capsule around a great deal trying to find them, and this was a costly bit of travel, too. It is possible to change the capsule's attitude gradually just by setting up a gentle movement with the controls and then letting it drift into the desired position. This is the most economical way to do it; you save fuel this way. But it is also a slow, time-consuming procedure and I was impatient to complete all of the items on the flight plan.

So I kicked the capsule around faster by using up more fuel and pushing it all the way. It was an expenditure that I would regret later on."[26] Three more flares were meant to be fired upwards on each of his second and third passes, but the experiment was discontinued when conditions over Woomera failed to improve.

Back in the Mercury Control Center, Gus Grissom and John Glenn follow the progress of the trouble-plagued MA-7 mission. (Photo: NASA)

Unlike Glenn on MA-6, who had taste-tested tubes of near-liquid food on his flight, one of Carpenter's tasks involved eating bite-sized chunks of compressed food to check that an astronaut could eat and swallow normally while weightless. Each chunk was coated with a glaze developed to keep the food from drying out and also to keep it from crumbling. There were pieces of orange peel and almond, high protein cereal and almond, cereal and raisins, a peanut cream bar, a compressed date nut bar, milk chocolate and chunks of rye bread. While in touch with the Woomera tracking station, Carpenter reported that he took four swallows of water and to his annoyance his food tablets had crumbled in their plastic bag. "Every time I opened the bag, some crumbs would come floating out, but once a bite-sized piece of food was in my mouth, there was no problem."[27]

After leaving Woomera behind, Carpenter put his spacecraft into a drifting mode to help conserve precious fuel. He was somewhat amused to find that by rocking his arms back and forth, like attempting a full twist on a trampoline, he could actually cause *Aurora 7* to react in all three axes – pitch, roll and yaw.

Continuing with his program, Carpenter then began his observations of the night sky. He commented on the fact that even though the stars were bright, he believed he saw less than he would observe on a clear night on the ground (again due to the lights in his cabin). However, he found that he could track them quite easily and hold *Aurora 7* in the correct attitude by fixing on a known star on the horizon and keeping it centered in the window.

When a metal washer suddenly appeared out of nowhere and began floating around inside the cabin, he plucked it out of the air and wondered which of the fittings it had come from.

Positioned to the right side of his helmet was a glass globe filled with fluids. Designed by the Lewis Research Center, he was to monitor the experiment, observing and taking note of any movements of the fluids during weightless flight (described in his post-flight report). As well, using a specially modified 35 mm camera fitted with UV-17 filters and a capacity for taking around 250 photographs, he snapped around 60 horizon definition photographs for the Massachusetts Institute of Technology to assist in the development of a navigation system to guide astronauts back to Earth from lunar flights. He also took photographs of cloud cover and other weather phenomena at the request of the Weather Bureau. He was so busy taking photographs that his other scheduled tasks inexorably began to slip behind the established timetable.

JOHN GLENN'S "FIREFLIES"

Just over an hour into the flight, still in drifting mode, Carpenter passed over the Canton Island station and checked his attitude readings with telemetry. He would soon experience his first sunrise, and yet another previously reported phenomenon would cause him a major distraction.

"Then, as the Sun rose ahead of me, I got my first look at John Glenn's 'fireflies.' As they drifted across the capsule near the window they looked more like snowflakes to me, whitish in color, and varying in size from one-sixteenth to one-half an inch in diameter."[28] He would observe and photograph this phenomenon that had really excited his interest before the flight. "I believe that they reflected sunlight and were not truly luminous [as reported by Glenn]," he stated in his post-flight report.

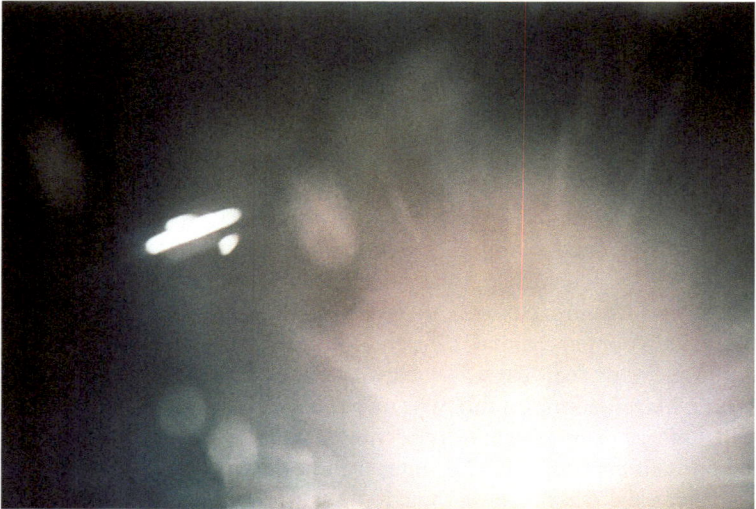

The then-mysterious "firelies" reported by John Glenn were photographed by Carpenter against the rising Sun. (Photo: NASA)

By this time he was feeling decidedly hot and uncomfortable in his tight-fitting space suit. Weightless sweat on his face was getting into his eyes, making it difficult to see clearly. He tried different settings on the dial on his suit that was meant to control the temperature within the suit, while ground stations offered up suggestions for bringing the temperature down, but nothing seemed to be working as it ought to. He would later say that it was not so much the suit temperature – which never registered more than 84°F – but the humidity that was causing the main problem. He found some relief by raising the visor on his helmet to expose his face to the hotter but drier atmosphere of the cabin. He found this to be far more comfortable, but once again he discovered that dealing with this problem had set him even further back in his flight schedule.

Then, passing below him was the United States. As he recalled later, "In areas not covered by clouds I could see the ground remarkably well, note lakes and rivers, and as I passed over farm country in the southwest I could even see places where the 'south 40' was cultivated and the 'north 40' was lying fallow. At every new sight my elation was renewed. I could hardly wait to get to the next one."

As he flew over Cape Canaveral at the end of his first orbit, Carpenter checked his gauges, and was "appalled" at the low state of his fuel. The readings were 69 percent capacity for the capsule's manual fuel, and the same reading for his automatic fuel. Not critical at this time, but a cause for some concern.

"I had not much more than half my supply left for both the automatic and manual control systems. I was warned that if I didn't conserve fuel, I would have to come down at the end of the second orbit. Cutting the flight short sounded like a terrible prospect to me. There was so much to be seen and done that I needed all the orbits I could get. I made up my mind to be very careful with the fuel."[29]

LAUNCHING A BALLOON

According to NASA's post-launch memorandum report, at 1 hour and 38 minutes into his flight, Carpenter initiated a major scheduled experiment when he deployed the tethered balloon by firing an activating squib. A small compressed spring then ejected the balloon and an inflatable bottle from a cylindrical container located in the antenna canister, along with two balsa block liners and the Mylar "confetti" discs. The balsa blocks were semi-cylindrical in shape, about six inches long and three inches wide. One was coated Day-Glo orange and black and the other was Day-Glo yellow and black. The quarter-inch Mylar discs, placed in the folds of the balloon to be dispersed when the balloon was deployed, were coated with aluminum foil on one side and a diffuse reflecting material on the other.

The balloon was tethered to *Aurora 7* by a 6-pound-test nylon line 100 feet in length, which was deployed from a spinning reel. When the balloon had been fully deployed, the line was entirely stripped from the reel, but it remained attached to a small strain gauge in the bottom of the balloon container. Continuous strain gauge measurements were to be recorded onboard the spacecraft until the drag test was completed, whereupon the balloon would be jettisoned and the rate and distance of separation between the spacecraft and balloon were to be estimated by Carpenter.[30] Unfortunately, the balloon did not inflate completely (the fault was later attributed to a ruptured seam in the skin) and it did not jettison properly. This meant that the precise drag measurements while tethered and subsequent visual estimates of the rate of separation after being jettisoned were not able to be obtained.

At deployment, Carpenter reported seeing the Mylar discs spread out and soon disappear. His first impression was that the balloon had broken loose from *Aurora 7*, but the object he was tracking was actually one of the balsa blocks.

"I have only the rectangular shape tumbling at this point about 200 yards back, barely visible; and now wait, here is a line. That was the cover, the balloon is out."

He observed the balsa block for about 20 seconds, at which time the balloon came into view, but as he later recorded in his pilot's flight report, the inflation process had not gone according to plan.

"Finally, the balloon came into view; it looked to me like it was a wrinkled sphere about 8 to 10 inches thick. It had small protrusions coming out each side. The balloon motion following deployment was completely random."[31]

"Balloon deploy, now," Carpenter reported. "The balloon is out and off. I see it way out, but it – I think now it is way out, and drifting steadily away. I don't see the line. I don't see that any attempt was made to inflate the thing. It's just drifting off."

While passing over the Bermuda tracking station five minutes later, he would confirm a further problem with the balloon experiment.

"I have lost sight of the balloon at this time .... Also, Bermuda, the balloon not only oscillates in cones in pitch and yaw, it also seems to oscillate in and out toward the capsule; and sometimes the line will be taut, other times it's quite loose."

Disappointed that the experiment had not worked as planned, Carpenter would continue to monitor the balloon over the next two orbits before attempting to jettison it ahead of the re-entry process. However, while the balloon was deployed, a series of spacecraft maneuvers seemed to have fouled the tethering line on the destabilizing flap located on the end of the cylindrical portion of the spacecraft, preventing the jettisoning of the balloon. "I continued to trail the balloon until retrofire," he later observed, "like a tin can attached to the rear bumper of a car."[32]

In order to conserve precious fuel, Carpenter set *Aurora 7* into drifting mode. He would start the spacecraft moving in one direction, and then simply allow it to continue to slowly rotate in that direction.

"I was able to make many observations that way, and I was impressed again and again with the wonder of weightlessness," he later told *Life* magazine. "A change of attitude means nothing in this state. Everything floats. Nothing rises or falls. 'Up' loses all significance. You can assign your own 'up' and put it anywhere – toward the ground, toward the horizon, or on a line drawn between two stars – and it is perfectly satisfactory. At one period I spent some time just playing with the camera, bouncing it off the fingertips of one hand and stopping it with the fingers of the other. Then I started it spinning slowly around in the air in front of me."[33]

With one orbit gone and two more to go, Carpenter had less than three hours remaining to complete a tightly packed program of experiments and observations while constantly battling rapidly diminishing fuel, a spacecraft beset with unresolved attitude issues, and compounding all of this, an overheating pressure suit. The remaining hours of the flight would severely test the mettle of the man, with many unexpected challenges to face, and an unrelenting timetable.

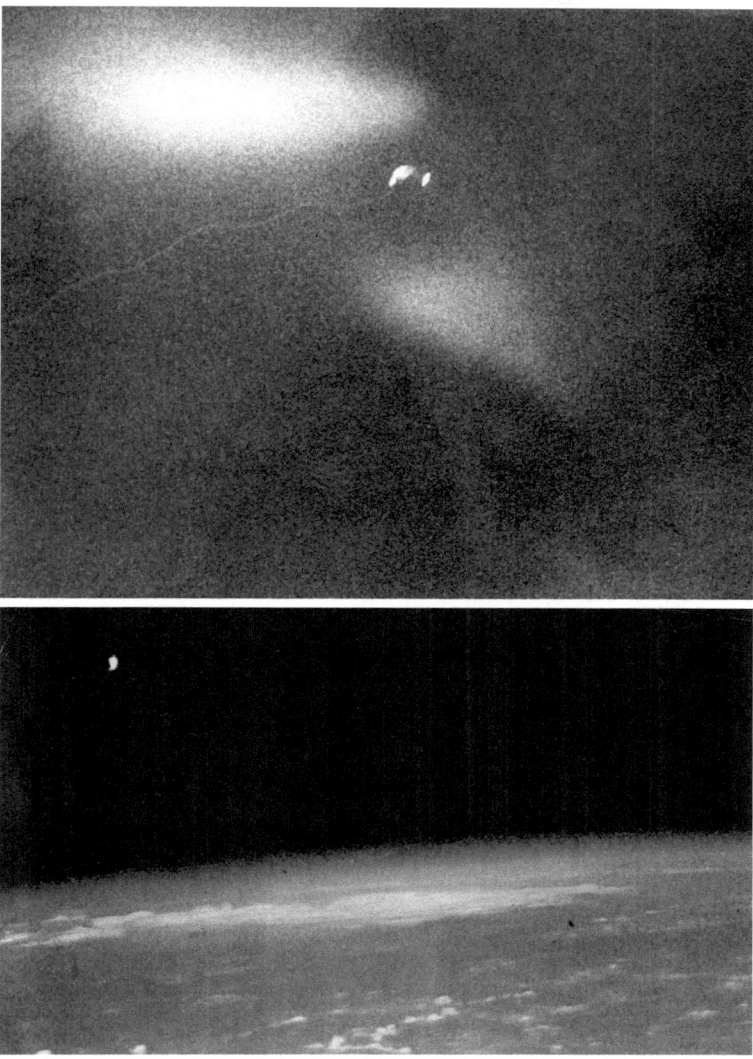

A major disappointment on the MA-7 flight was the failure of the balloon experiment, seen here trailing behind *Aurora 7*. (Photos: NASA)

REFERENCES

1. Scott Carpenter, *Life* magazine article, "I got let in on the great secret," issue 8 June 1962, pg. 30
2. *Ibid*
3. Eddie Pugh and Nigel Macknight, *Space Flight News* magazine article, "Flashback Mercury 7," issue No. 52, April 1990, pg. 34
4. Francis French and Colin Burgess, *Into That Silent Sea: Pathfinders of the Space Era, 1961–1965*, University of Nebraska Press, Lincoln, NE, 2007

5. *Ibid*
6. Scott Carpenter, *Life* magazine article, "I got let in on the great secret," issue 8 June 1962, pg. 30
7. Rene Carpenter, *Life* magazine article, "Scott Carpenter and his son and his wife living through 'the time that grew too long'," issue 1 June 1962, pg. 30
8. Scott Carpenter, *Life* magazine article, "I got let in on the great secret," issue 8 June 1962, pg. 30
9. *Ibid*, pg. 32
10. NASA *Space News Roundup*, "MA-7 Conference Reviews Results in Detail," Vol. 1, No. 22, 22 August 1962, pg. 1
11. *Ibid*
12. Scott Carpenter, *Life* magazine article, "I got let in on the great secret," issue 8 June 1962, pg. 30
13. M. Scott Carpenter, L. Gordon Cooper, Jr., John H. Glenn, Jr., Virgil I. Grissom, Walter M. Schirra, Jr., Alan B. Shepard, Jr., Donald K. Slayton, *We Seven*, Simon and Schuster, Inc., New York, NY, 1962, pg. 448
14. Scott Carpenter and Kris Stoever, *For Spacious Skies: The Uncommon Journey of a Mercury Astronaut*, Harcourt, Orlando, FL, 2002
15. M. Scott Carpenter, "Pilot's Flight Report," from *Results of the Second U.S. Manned Space Flight, May 24, 1962*, NASA publication SP-6, 1962
16. M. Scott Carpenter, L. Gordon Cooper, Jr., John H. Glenn, Jr., Virgil I. Grissom, Walter M. Schirra, Jr., Alan B. Shepard, Jr., Donald K. Slayton, *We Seven*, Simon and Schuster, Inc., New York, NY, 1962, pg. 449
17. Scott Carpenter and Kris Stoever, *For Spacious Skies: The Uncommon Journey of a Mercury Astronaut*, Harcourt, Orlando, FL, 2002
18. M. Scott Carpenter, L. Gordon Cooper, Jr., John H. Glenn, Jr., Virgil I. Grissom, Walter M. Schirra, Jr., Alan B. Shepard, Jr., Donald K. Slayton, *We Seven*, Simon and Schuster, Inc., New York, NY, 1962, pg. 450
19. M. Scott Carpenter, "Pilot's Flight Report," from *Results of the Second U.S. Manned Orbital Space Flight, May 24, 1962*, NASA SP-6, MSC, Houston, TX, 1962
20. Scott Carpenter and Kris Stoever, *For Spacious Skies: The Uncommon Journey of a Mercury Astronaut*, Harcourt, Orlando, FL, 2002
21. *Ibid*
22. Gene Kranz, *Failure Is Not an Option*, Simon and Schuster, New York, NY, 2000
23. Scott Carpenter and Kris Stoever, *For Spacious Skies: The Uncommon Journey of a Mercury Astronaut*, Harcourt, Orlando, FL, 2002
24. Scott Carpenter, *Life* magazine article, "I got let in on the great secret," issue 8 June 1962, pg. 32
25. M. Scott Carpenter, L. Gordon Cooper, Jr., John H. Glenn, Jr., Virgil I. Grissom, Walter M. Schirra, Jr., Alan B. Shepard, Jr., Donald K. Slayton, *We Seven*, Simon and Schuster, Inc., New York, NY, 1962, pp. 450–451
26. *Ibid*
27. *Results of the Second U.S. Manned Orbital Space Flight, May 24, 1962*, NASA Publication SP-6, Manned Spacecraft Center, Houston, TX
28. M. Scott Carpenter, L. Gordon Cooper, Jr., John H. Glenn, Jr., Virgil I. Grissom, Walter M. Schirra, Jr., Alan B. Shepard, Jr., Donald K. Slayton, *We Seven*, Simon and Schuster, Inc., New York, NY, 1962, pg. 452
29. Scott Carpenter, *Life* magazine article, "I got let in on the great secret," issue 8 June 1962, pg. 32
30. *Post-launch Memorandum Report for Mercury-Atlas No. 7 (MA-7), Part 1: Mission Analysis*, NASA Manned Spacecraft Center/Cape Canaveral, Florida, 15 June 1962
31. *Results of the Second U.S. Manned Orbital Space Flight, May 24, 1962*, NASA SP-6, Manned Spacecraft Center, Houston, TX
32. Scott Carpenter, *Life* magazine article, "I got let in on the great secret," issue 8 June 1962, pg. 32
33. *Ibid*

5

A highly troubled mission

"After the initial sensation of weightlessness, it was exactly what I had expected from my brief experience with it in training. It was very pleasant, a great freedom, and I adapted to it quickly. Movement in the pressure suit was easier and the couch was more comfortable."[1]

Despite the time Scott Carpenter devoted to flying *Aurora 7* and monitoring the spacecraft's many systems, he also managed to carry out a vast number of mission tasks. Just as he had trained hard in preparing for the operational side of the MA-7 mission, he had also taken a keen interest in the science experiments he would conduct during his flight, and was making his way through a prodigious number of authorized tests and trials for other agencies and institutions. Had he not been encumbered by systems failures within the spacecraft, his heavy, self-imposed workload would have been the source of endless compliments.

Just above his right ear hung the glass globe in which he studied the movements of fluids during weightlessness for the NASA Lewis Research Center. Using a photometer he measured the brightness of stars and other outside objects.[2] As well, using special filters he would eventually take 60 photographs of the Earth's horizon for the Massachusetts Institute of Technology, which wanted them for the development of a navigation system to guide astronauts returning to Earth from lunar expeditions.

SECOND ORBIT

As he passed over the Canary Islands on his second orbit, Scott Carpenter received yet another warning about his fuel levels.

01 50 15.5 (Canary Islands): *Aurora Seven*, you are fading rapidly. You are fading. MCC [Mercury Control Center] is worried about your auto fuel and manual fuel consumption. They recommend that you try to conserve your fuel.

01 50 28.5 (Pilot): Roger. Tell them I am concerned also. I will try and conserve fuel.

Kano, Nigeria was the next station along his orbital track. During their communications Carpenter reported on his physical condition and that of his spacecraft.

© Springer International Publishing Switzerland 2016

C. Burgess, *Aurora 7*, Springer Praxis Books, DOI 10.1007/978-3-319-20439-0_5

Water-damaged footage of Scott Carpenter during MA-7, showing the fluid movement experiment left of photo above his right shoulder. (Photo: NASA)

The Canary Islands tracking station. (Photo: Grand Canary Islands website, unaccredited photo)

01 55 08.5 (Pilot): Roger. My status is good; fuel reads 51 [percent] and 69 [percent]; oxygen is 84 [percent] and 100 [percent]; cabin pressure is holding good. All DC and AC power is good. The only thing to report regarding the flight plan is that fuel levels are lower than expected. My control mode now is ASCS. I expended my extra fuel in trying to orient after the night side. I think this is due to conflicting requirements of the flight plan. I should have taken time to orient and then work with other items. I think that by remaining in automatic, I can keep … stop this excessive fuel consumption. And the balloon is sometimes visible and sometimes not visible. I haven't any idea where it is now, and there doesn't seem to … it seems to wander with abandon back and forth, and that's all, Kano.

Kano acknowledged this message and asked how he was feeling; his body temperature appeared a little high.

01 57 07.5 (Pilot): Roger. I feel fine. Last time around I … someone told me it was 102 [degrees]. I don't feel, you know, like I'm that hot. Cabin temperature is 101 [degrees]. I'm reading 101 [degrees], and the suit temperature indicates 74 [degrees].

When asked if he was perspiring, he gave confirmation of this.

0t 57 41.5 (Pilot): Slightly, on my forehead. Since turning down the suit water valve, the suit steam vent temperature has climbed slightly … am increasing from one to two at this time. This should bring it down. The cabin steam vent temperature has built back up to 40 [degrees].

Kano indicated they were reading his body temperature at 102 degrees at that time, but he confirmed once again that he was feeling fine. His next contact station was the Indian Ocean ship *Coastal Sentry*. Carpenter told them he would check his fly-by-wire in order to diagnose any problem with the hydrogen peroxide thrusters, as he believed he might have some automatic mode difficulty. Soon thereafter he advised that all his thrusters were okay.

02 09 17.5 (Pilot): However, the gyros do not seem to be indicating properly … And that is not correct either. The gyros are okay; but on ASCS standby. It may be an orientation problem. I'll orient visually and . . . see if that will help out the ASCS problem.

Once again he reported ongoing problems with the steam vent temperature in his suit. He decided to try opening the visor, and told Kano that even with the high cabin temperature it seemed cooler with the visor opened. As he later related, "The dry air would at least provide some evaporative relief from the sweat now pouring down my forehead, plowing through my eyebrows, and stinging my eyes with salt."[3] He then mentioned that with all the systems and other problems, he had gotten behind in the flight plan.

In his post-flight pilot's report, Carpenter said that he noticed very little sensation of motion or movement during the flight. "My only clues to motion were the instruments and the view through the window and periscope. At times during the flight, the spacecraft angular rates were greater than 6 degrees per second, but aside from vision, I had no sense of movement. I was never disoriented. I always knew where the controls and other objects were relative to myself. I could reach anything I needed. I did have one unusual experience. After looking out the window for some time, I noticed that when I turned my head to the right to look at the special equipment storage kit, I would get the impression that it was oriented vertically, or 90 degrees from where I felt it should be. This impression was because of my training in the procedures trainer and lasted only temporarily.

"At times when the gyros were caged and nothing was visible out the window, I had no idea where the Earth was in relation to the spacecraft. However, it did not seem important to me. I knew at all times that I only had to wait and the Earth would again appear in the window. The periscope was particularly useful in this respect, because it had such a wide field of view. Even without it, however, the window would have been adequate."[4]

Despite ongoing problems with his spacecraft attitude, Carpenter continued taking photos of the beautiful Earth below. (Photo: NASA)

FLIGHT CONTINUES

Passing over Muchea once again, Carpenter reported to CapCom Deke Slayton that his status was good and he was on fly-by-wire mode, but he was still concerned about his inability to lower the temperature within his pressure suit. Slayton mentioned that the capsule's cabin temperature was also excessively high.

After eating some more morsels of food, Carpenter continued to try to regulate the temperature within his suit, but to little avail. As he later reported to the Woomera station, his suit temperature was a little better at 74 degrees – some two degrees lower – and the suit's steam exhaust was 71 degrees. He was asked if he was feeling a little more comfortable at that time.

02 34 02.5 (Pilot): I don't know. I'm still warm and still perspiring, but not really uncomfortable. I would like to … I would like to nail this suit temperature problem down. It … for all practical purposes, it's uncontrollable as far as I can see.

Following further discussion on the problem of his suit temperature, Woomera asked if he had eaten any food.

02 34 53 (Pilot): Yes, I have. However, the food has crumbled badly; and I hate to open the package any more for fear of getting crumbs all over the capsule. I can verify that eating bite-size food as we packaged for this flight is no problem at all. Even the crumbly foods are eaten with no … with no problem.

02 35 20 (Woomera): Roger. How about water?

02 35 22.5 (Pilot): I had taken four swallows at approximately this time last orbit. As soon as I get the suit temperature pegged a little bit, I'll open the visor and have some more water. Over.

The Canton Island station asked whether he had any specific observations on the balloon experiment, and if he could report which of the five colors was most visible. Although the balloon was only partially inflated and was oscillating behind *Aurora 7*, Carpenter reported that in his opinion the Day-Glo orange stood out most.

At this point in the flight, Carpenter's second sunrise was only three to four minutes away, and he asked the next station, Hawaii, to give him some time before reporting his status as he had other important duties to perform. "Sunrises and sunsets were extremely busy time-blocks during Mercury flights," he later noted. "There were important measurements to make of the airglow and other celestial phenomena and innumerable photographs to take."[5] He was also observing some more of the strange translucent flakes out of his window. But Hawaii was insistent on first having a short report on his condition, and also suggested that on medical advice he should drink some more water. He would later admit that he should probably have consumed more water during the flight to prevent a possible problem with dehydration, but there were numerous technical problems and a steadily slipping work schedule to distract him.

His fuel report indicated that the spacecraft's manual fuel supply was now down to 45 percent and his automatic supply stood at 62 percent. When asked if he had an auto-fuel warning light his response appeared to be a little impatient with the question.

02 50 48 (Pilot): That is right. I have reported it, and I believe I reported it a long time ago. It is covered with tape at the moment.

The Hawaii CapCom persisted in pressing Carpenter for status information, but even though he complied, he continued to be more concerned with reporting other events taking place outside his window.

02 52 20.5 (Hawaii): *Aurora Seven*. This is Cap Com. We'd like for you to return to gyros normal and see what kind of indication we have; whether or not your window view agrees with your gyros.

02 52 34 (Pilot): Roger. Wait one.

02 52 47 (Pilot): I have some more of the white particles in view below the capsule. They appear to be traveling exactly my speed. There is one drifting off. It's going faster than I am as a matter of fact.

02 53 11.5 (Hawaii): Roger. Understand.

02 53 15 (Pilot): I haven't seen the great numbers of these particles, but I've seen a few of them. Their motion is random; they look exactly like snowflakes to me.

02 53 29 (Hawaii): Roger. Have you tried returning ....

02 53 33 (Pilot): Negative. Let me get within scanner limits first.

02 53 39 (Hawaii): Say again.

02 53 40 (Pilot): I must adjust my attitude to within scanner limits first.

02 53 46.5 (Hawaii): Roger.

02 54 18.5 (Pilot): There were some more of those little particles. They definitely look like snowflakes this time.

02 54 26 (Hawaii): Roger. Understand. Your particles look like definite snowflakes.

02 54 32 (Pilot): However …

02 54 33.5 (Hawaii): Can we get a blood pressure from you, Scott?

02 54 34.5 (Pilot): Roger. Blood pressure …start … now. I have the balloon …now … pretty steadily below me. Not oscillating. And go to gyros normal. Gyros normal now.

"GO" FOR A THIRD ORBIT

At this point Hawaii faded from contact, and the next station on his track was Point Arguello, California, manned by CapCom Alan Shepard who, after confirming solid communications, asked for a short report. Carpenter replied that his fuel levels were now down to 45 percent in the manual supply and 50 percent in the automatic supply. He also stated that his suit temperature had dropped a little further, and now stood at 70 degrees. There was some good news in store when Shepard confirmed that the MA-7 mission had been given the "go" to complete a third orbit. However there was a further warning from Mercury Control about his excess fuel consumption.

03 00 15 (Point Arguello): General Kraft is still somewhat concerned about auto fuel. Use as little auto … use no auto fuel unless you have to prior to retrosequence time. And I think maybe you might increase flow to your inverter heat exchanger to try to bring the temperature down. They are not critical yet, however.

As Carpenter continued his report to Shepard, he was still keen to comment on the flakes he could see outside the window.

03 01 35 (Pilot): All right now, I'm beginning to get all of those various particles, they … they're way out. I can see some that are a 100 feet out.

03 01 52 5 (Point Arguello): Roger. Real [far] off.

03 01 55.5 (Pilot): They all look like snowflakes to me. No don't … they do not glow of their own accord.

Shepard, however, was keen to get further information on his colleague's well-being in the short time they were in communication.

03 02 12 (Point Arguello): Roger, *Seven*. Do you …have you …stopped perspiring at the moment?

03 02 20 (Pilot): No, I'm still perspiring, Al. I think I'll open up the visor and take a drink of water.

03 02 27 (Point Arguello): Roger. Sounds like a good idea.

03 02 42 (Point Arguello): *Seven*, would you give us a blood pressure, please, in between swallows?

03 03 27 (Pilot): Okay, there's your blood pressure. I took about 20 swallows of water. Tasted pretty good.

Two orbits down, and one to go. As the spacecraft swept over Cape Canaveral on its third pass, Carpenter was in voice contact with CapCom Gus Grissom. After giving a status check, he was asked to expand upon the actions of the balloon trailing behind *Aurora 7*.

03 08 35 (Pilot): Yes, it has a random drift. There is no oscillation that I can predict whatsoever. The … the line leading to the balloon sometimes is tight; sometimes is loose enough, so that there are loops in it. Its behavior is strictly random as far as I can tell. The balloon is not inflated well either. It's an oblong shape out there, rather than a round figure; and I believe when the Sun is on it, the Day-Glo orange is the most brilliant, and the silver. That's about all I can tell you, Gus.

03 09 28.5 (Cape Canaveral): Roger. Surgeon suggests that you drink as much water as you can. Drink it as often as you can.

03 09 38.5 (Pilot): Roger.

Gus Grissom and John Glenn in Mercury Control with (center) Flight Director Chris Kraft. (Photo: NASA)

Later in the conversation, Grissom asked Carpenter if he had done any drifting flight, in order to conserve his dwindling fuel.

03 11 38.5 (Pilot): That is roger. And if I am to save fuel for retrosequence, I think I better start again. Over.

03 11 49 (Cape Canaveral): Roger, I agree with you.

03 11 52 (Pilot): My control mode is now manual; gyros are caged, and I will allow the capsule to drift for a little while.

Further discussion on the "snowflakes" and the failed balloon experiment followed. With only three or four minutes of communication time left, and while Carpenter was scheduled to jettison the balloon ahead of re-entry, Grissom repeated the fuel-status admonition from Mercury Control.

Exterior view of the Mercury Control Center and press site, and controllers at their stations within the building. (Photos: NASA)

03 14 26.5 (Cape Canaveral): We're still fairly happy with your fuel state now. Don't let … we'd like for you not to let either get down below 40 percent.

03 14 33 (Pilot): Roger. I'll try. I have balloon jettison on and off, and I can't get rid of it.

03 14 41 (Cape Canaveral): Understand that you can't get rid of the balloon.

03 14 43.5 (Pilot): That's right. It will not jettison.

03 14 48.5 (Cape Canaveral): Okay.

Carpenter then gave a further description of the actions of the particles outside the window, reporting that they were still streaming aft in an arc of around 120 to 139 degrees. With communication almost at an end, Grissom asked for any information on the zero-g fluid experiment.

03 16 53.5 (Pilot): Roger. At this moment, the fluid is all gathered around the stand-pipe; the standpipe appears to be full and the fluid outside the standpipe is about halfway up. There is a rather large meniscus. I'd say about 60 degree meniscus.

DRIFTING FLIGHT

Communication with the Cape was failing at this time, as *Aurora 7* swept on, still in drifting mode, ready for a third pass over the Canary Islands ahead. "Drifting flight was effortless and created no problems," he would state in his post-flight pilot's report. "Aligning the gyros consumed fuel or time. The horizon provided a good roll and pitch reference as long as it was visible in the window. On the dark side of the Earth, the horizon or the airglow layer is visible at all times, even before moonrise. Yaw reference was a problem. The best yaw reference was obtained by pitching down minus 50 to minus 70 degrees and looking through the window. The periscope provided another good yaw reference at nearly any attitude. The zero-pitch mark on the periscope was also a valuable reference for aiming the gyros since at zero pitch, the horizon could not be seen through the window. Yaw attitude is difficult to determine at night, and the periscope is of little help in determining yaw on the night side. The best reference is a known star."[6]

In the Mercury Control Center, Assistant Flight Director Gene Kranz was closely monitoring the continuing problems with the ASCS and the resultant excess fuel usage, and there were deep concerns, as he later reflected:

"The Mercury capsule design provided the astronaut with two attitude control systems – an automatic system containing 12 thrusters and 32 pounds of fuel (hydrogen peroxide). In Mercury Control, [flight controller] Arnie Aldrich was watching the drain of fuel from the tanks. Whether on the manual or automatic attitude control system, the high usage continued and, given the infrequent site contacts with the spacecraft, Aldrich was unable to identify the cause. Carpenter was repeatedly advised to conserve fuel by turning off all control and going into drifting flight. At the start of the third and final orbit the propellants were down to 45 percent remaining in both systems. [Chris] Kraft's concern at this time was not the fuel level as much as the control techniques used by Carpenter. Every time he maneuvered the capsule the fuel quantities plummeted. If the trend we were observing continued, Carpenter would run out of attitude control fuel before re-entry. Scott was again told to go to drifting flight and conserve his fuel for retrofire and entry. The site reports across Africa and Australia indicated that the fuel usage had stopped."[7]

Carpenter continued to pass on information of his well-being and the status of his on-board systems, and confirmed his readings as he swept over Kano. With less than an hour left before retrofire, the astronaut continued to conduct experiments, take photographs, and monitor his condition. *Aurora 7* then passed over the Indian Ocean ship, *Coastal Sentry* and he reported in.

03 39 31.5 (Pilot): Roger. My status is good, the capsule status is good. I am in drifting flight on manual control. Gyros are caged, the fuel reads 45 to 42 [percent], oxygen 79 to 100 [percent]. Steam vent temperatures both read 65 [degrees] now; suit temperature has gone down nicely. It is now 62 [degrees], and all the power is good. The blood pressure is starting at this time. I've just finished taking some MIT [airglow] pictures, and that is all I have to report at this time.

The USAF Missile Range Instrumentation ship *Coastal Sentry* (call sign *Coastal Sentry Quebec*) was used as a communication station for MA-7 while stationed off the south-eastern coast of Africa. (Photo: USAF)

He then added that he still had the spacecraft in a drifting mode, but remained unable to jettison the balloon. The CapCom stated that this should present no problem during the re-entry phase. The line attaching it to *Aurora 7* would simply burn through and detach very early in the process. For the record, Carpenter then reported on the glorious colors associated with the sunset he could see out the window.

03 43 20 (Pilot): The sunsets are most spectacular. The Earth is black after the Sun has set. The Earth is black; the first band close to the Earth is red, the next is yellow; the next is blue; the next is green; and the next is sort of a … sort of a purple. It's almost like a very brilliant rainbow.

He was then asked to confirm once again that the balloon was still attached, which he confirmed. After a few minutes, and before doing an exercise in trying to disorientate his senses, he decided it was time to relax for a few moments, stretch out, and drink some more water.

The balloon, still tagging along behind *Aurora 7*. (Photo: NASA)

03 48 50 (Pilot): At 3 hours and 48 minutes and 51 seconds elapsed, I'm taking a good swig of water. It's pretty cool this time. Stretching my legs a tad. It's quite dark. I'm in drifting flight. Oh, boy! It feels good to get that leg stretched out. That one, and the right one too.

03 51 13.5 (Pilot): Okay. I'm shaking my head violently from all sides, with eyes closed, up and down, pitch, roll, yaw. Nothing in my stomach; nothing anywhere. There is now … I will try to poke zero, time zero button. Well, I missed it. I was a little disoriented … as to exactly where things are; not sure exactly what you want to accomplish by this but there is no problem of orienting. Your … inner ears and your mental appraisal of horizontal, you just adapt to this environment, like … like you were born in it. It's a great, great freedom.

SOLVING THE "SNOWFLAKES"

One of his reports to the ground at around that time indicated that the lamination on the star chart was far too shiny, making it difficult to read. All feedback was important, as changes could then be incorporated into future missions.

As the flight continued, Carpenter reported on his condition, his experiments, the status of his spacecraft and fuel and comfort levels. Upon passing over Woomera for the final time, he believed he had finally solved the mystery of John Glenn's "fireflies."

04 19 22.5 (Pilot): Sunrise. Ahhhhh! Beautiful lighted fireflies that time. It was luminous that time. But it's only, okay, they … all right, I have … if anybody reads, I have the fireflies. They are very bright. They are capsule emanating. I can rap the hatch and stir off hundreds of them. Rap the side of the capsule; huge streams come out. They … some appear to glow. Let me yaw around the other way.

04 20 25 (Pilot): Some appear to glow but I don't believe they really do; it's just the light of the Sun. I'll try to get a picture of it. They're brilliant. I think they would really shine through on the photometer. I'll rap. Let's see.

04 21 39.5 (Pilot): Taking some pictures at F/2.8 and bulb. The pictures now, here, one of the balloon. The Sun is too bright now. That's where they come from. They are little tiny white pieces of frost. I judge from this that the whole side of the capsule must have frost on it.

Although later communications between *Aurora 7* and the tracking station at Hawaii were weak, Carpenter heard enough to know that his next task was to reorient his attitude and go onto autopilot. Time-wise, however, things were getting really tight, and he was mindful of the fact that he still had to stow a lot of his loose equipment. The flight plan called for him to tend to the retrosequence checklists at 04 24 00. Beyond that, it was only another eight and a half minutes until retrofire. Hawaii began urging things along. When the time came to begin the retrosequence checklists the station knew they had only a few minutes remaining before they would lose communication with *Aurora 7*. Their messages became increasingly urgent.

04 24 20 (Hawaii): Roger. Are you ready to start your pre-retrosequence checklist?

04 24 23.5 (Pilot): Roger. One moment. I'm aligning my attitudes. Everything is fine. I have part of the stowage checklist taken care of at this time.

04 25 11.5 (Hawaii): *Aurora Seven*, do you wish me to read out any of the checklist to you?

04 25 17 (Pilot): Let me get the stowage and then you can help me with the pre-retrograde.

04 25 24 (Hawaii): Standing by.

04 25 55 (Hawaii): *Aurora Seven*, can we get on with the checklist? We have approximately 3 minutes left of contact.

04 26 00 (Pilot): Roger. Go ahead with the checklist. I'm coming to retroattitude now and my control mode is automatic and my attitudes … standby. Wait a minute, I have a problem in … I have an ASCS problem here. I think ASCS is not operating properly. Let me … Emergency retrosequence is armed and retro manual is armed. I've got to evaluate this retro … this ASCS problem, Jim, before we go any further.

Two minutes later, after Carpenter informed them that his emergency drogue deploy and emergency main fuses were off and that he was going back onto fly-by-wire, Hawaii became increasingly insistent, even as communications began to become erratic.

04 28 06 (Hawaii): Scott, let's try and get some of this retrosequence list checked off before you get to California.

04 28 12.5 (Pilot): Okay. Go through it, Jim. Jim, go through the checklist for me.

Within seconds of the retrosequence checklist being completed, *Aurora 7* flew beyond range of the Hawaii station. It had been a close call.

In a frank admission in his post-flight pilot's report, Carpenter states that one reason he later got behind at retrofire was because, just at dawn, he believed he had discovered the source of the space "snowflakes" or "fireflies" which had earlier puzzled John Glenn. "A number of times during the flight, I observed the particles reported by John Glenn. They appeared to be like snowflakes. I believe that they reflected sunlight and were not truly luminous. The particles traveled at different speeds, but they did not move away from the vehicle as rapidly as the confetti that was deployed upon balloon release. At dawn on the third orbit as I reached for the densitometer, I inadvertently hit the space-craft hatch and a cloud of particles flew by the window. Since I was yawed to the right, the particles traveled across the front of the window from the right to the left. I continued to knock on the hatch and other portions of the spacecraft walls, and each time a cloud of particles came past the window. The particles varied in size, brightness, and color. Some were gray and others were white. The largest were four or five times the size of the smaller ones. One that I saw was a half-inch long. It was shaped like a cuticle and looked like a lathe turning.

"I felt that I had time to get that taken care of and still prepare properly for retrofire, but time slipped away. The Hawaii CapCom was trying very hard to get me to do the pre-retrograde checklist. After observing the particles, I was busy trying to get aligned in orbit attitude. Then I had to evaluate the problem in the automatic control system. I got behind and had to stow things haphazardly."[8]

Soon enough, Carpenter was back in touch with Alan Shepard at the Point Arguello station, who asked him if he was in retroattitude.

04 31 53 (Pilot): Yes, I don't have agreement with ASCS in the window, Al. I think I'm going to have to go to fly-by-wire and use the window and the scope. ASCS is bad. I'm on fly-by-wire and manual.

04 32 06 (Point Arguello): Roger. We concur. About 30 seconds to go …. About 10 seconds on my mark.

04 32 23.5 (Pilot): Roger.

04 32 28 (Point Arguello): 6, 5, 4, 3, 2, 1.

04 32 36 (Pilot): Retrosequence is green.

Shepard then asked Carpenter to do a quick check to see if the orientation mode would hold, and to acknowledge that if the gyros were off he would require to use attitude bypass. Carpenter said the gyros were off and he would use the attitude bypass and manual override. Shepard then began the countdown to retrofire.

04 33 00 (Point Arguello): 4, 3, 2, 1, zero.

RETROFIRE

"The rockets were supposed to fire automatically," Carpenter later recorded in *For Spacious Skies*. "I watched the second hand pass the mark, and when they didn't, punched the retro button myself a second later. An agonizing three seconds passed until the reassuring sound and vibration of firing retrorockets filled the cabin. I was prepared for a big boot, which never came. Deceleration was just a very gentle nudge, not at all the terrific push back toward Hawaii that John had reported feeling from his own retrofire. Al was still in voice range and we continued to transmit information about retrosequence. I noticed smoke in the cabin and the smell of metal. Two fuses had overheated. I was worried about the delayed firing of the retrorockets. At that speed, a lapse of three seconds would make me at least fifteen miles long in the recovery area. Al asked me if my attitudes held, and I said, 'I think they were good,' but I wasn't sure, adding that 'the gyros are not quite right.'"[9]

It would later be discovered that the three retrorockets had actually under-thrusted, which added another 60 miles to the capsule's overshoot.

The next phase of the re-entry process was the automatic jettison of the retropack. This was another anxious moment. At 04 34 10 he reported that retrojettison had occurred right on time, but that his fuel tanks were down to 20 and 5 percent respectively. Thirty seconds later he informed Shepard that the trailing balloon had disappeared from sight, but he was experiencing a tumbling motion that needed to be dampened. At this point he discovered he was completely out of manual fuel, and would have to go to fly-by-wire in order to stop the tumbling and get into the correct re-entry attitude.

"Retropack jettison and the retraction of the periscope occurred on time. At this time, I noticed my appalling fuel state and realized that I had controlled retrofire on both the manual and fly-by-wire systems. I tried both the manual and the rate-command control modes and got no response. The fuel gauge was reading about 6 percent, but the fuel tank was empty. This left me with about 15 percent on the automatic system to last out the [remaining] 10 minutes to 0.05 g and [then] to control the re-entry."[10]

04 36 29.5 (Point Arguello): How are you doing on re-entry attitude? Over.
04 36 32.5 (Pilot): Stowing a few things first. I don't know yet. Take a while.
04 36 46 (Pilot): Okay.
04 36 54 (Pilot): Going to be tight on fuel.
04 37 02.5 (Point Arguello): Roger. You have plenty of time; you have about 7 minutes before .05 g so take . . .
04 37 10 (Pilot): Roger.
04 37 28 (Pilot): Okay. I can make out very, very small … farm land, pasture land below. I see individual fields, rivers, lakes, roads, I think. I'll get back to re-entry attitude.
04 37 39.5 (Point Arguello): Roger *Seven*, recommend you get close to re-entry attitude using as little fuel as possible and stand by on fly-by-wire until rates develop. Over.
04 37 50 (Pilot): Roger. Will do.

Shepard then advised that they were losing communications with the spacecraft and asked Carpenter to stand by for contact with Gus Grissom at the Cape. When Grissom came online and communications checks were complete, he asked if Carpenter had his visor closed. This was confirmed, and then Grissom asked for a fuel reading.

04 41 20 (Pilot): Fuel is 15 [percent] auto. I'm indicating 7 [percent] manual, but it is empty and ineffective.

Grissom reminded Carpenter that there were only a few minutes until the beginning of the communications blackout during re-entry.

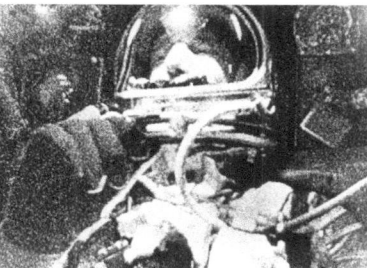

Carpenter prepares for retrofire. (Photo: NASA)

04 41 54 (Pilot): Roger. It's going to be real tight on fuel, Gus. I've got the horizon in view now. Trying to keep rates very low.

As Carpenter explained in his post-flight pilot's report, "I used [the automatic system] sparingly, trying to keep the horizon in the window so that I would have a correct attitude reference. I stayed on fly-by-wire until 0.05 g. At 0.05 g, I think I still had a reading of about 15 percent on the automatic fuel gauge. I used the window for attitude reference during re-entry because of the difficulty I had experienced with the attitude displays prior to retrofire."[11]

Grissom advised that the weather in the recovery area was good, with overcast cloud at 1,000 feet, 3-foot waves, 8 knots of wind, and 10 miles of visibility. He then pointed out that they were showing *Aurora 7* still had some manual fuel left, but Carpenter responded that he couldn't get anything out of it. As he recalled after the flight:

"In between the time I fired the retrorockets and the moment *Aurora 7* began its re-entry through the atmosphere, things were very tight indeed. My fuel supply was critically low and I was not at all sure that I had enough left to bring the capsule into the proper position. If we started the re-entry at the wrong angle and the fuel was exhausted, I would be unable to control the capsule during the descent. The chances that I would survive such an uncontrolled re-entry were not good.

"I have since heard recordings of my voice played at this tense time. I don't sound afraid or confused, but oddly dejected. All the elation so audible in earlier portions of my flight is gone. The tone of my voice has dropped measurably and my sentences all end on a minor key.

"But at this low point I remember thinking, 'This has been the greatest day of your life. You've got nobody to blame for being in this spot but yourself. If you do right, you may make it. If you don't do right, you're just going to buy the farm.' I decided I had to use more fuel to get into position for re-entry. It would have to last."[12]

At the time of retrofire, Carpenter believed he had brought *Aurora 7* to the correct attitude, but he said that he subsequently discovered this was incorrect.

"The small, bottle-top end of the capsule which was trailing me was canted 25 degrees to the right of where it should have been. In other words, I had an error in yaw. I had just not been able to line the capsule up on all three axes as precisely as I should have. This meant that the capsule was not pointed in an absolutely straight line along its path when the rockets fired, and so it did not slow down as much as it should have.

"If the capsule was lopsided during re-entry, one side of it could overheat, and the heat could even eat through the thin upper wall and destroy the parachutes that I would need to get down."[13]

COMING HOME

As *Aurora 7* began to re-enter the atmosphere, Carpenter dutifully noted his impressions on the on-board tape recorder. As he said in his post-flight pilot's report, "I began to hear the hissing outside the spacecraft that John Glenn had described. The spacecraft was aligned with 3 or 4 degrees in pitch and yaw at the start of the re-entry period. I feel that it would have re-entered properly without any attitude control. The gradual increase of aero-dynamic forces during the re-entry appeared to be sufficient to align the spacecraft properly."[14]

As it entered the atmosphere, a shockwave formed just in front of the spacecraft due to its tremendous speed. At temperatures in the vicinity of 3,000° Fahrenheit, this formed a plasma cloud around the capsule that blocked radio transmissions. This communications "blackout," as it became known, lasted about 12 minutes.

As *Aurora 7* plunged ever deeper into the thickening atmosphere Carpenter continued to record his impressions. Meanwhile in Mercury Control the tension was mounting with every passing minute because, with his depleted fuel supply, Carpenter could be re-entering unable to control the orientation of his capsule, and any tumbling motion could mean a horrible end for both the man and his machine.

After nine minutes of blackout, one of the monitoring instruments momentarily picked up the astronaut's heartbeat from sensors strapped to his body. But still there was no word from the man himself.

The flight was being broadcast across the nation. In New York's Grand Central Station terminal a hushed crowd had gathered by the huge television screen which earlier that day had broadcast the triumphant launch from Cape Canaveral. In the White House an anxious President Kennedy sat near a telephone that linked him directly with Mercury Control. He would be told immediately of any news – good or bad. In Mercury Control itself, CapCom Gus Grissom repeatedly called to establish contact with the spacecraft.

Shortly after the 0.05 g mark, Carpenter starting picking up oscillations on the pitch and yaw rate needles, much the same as those he had experienced in some of his training runs. From this, he deduced that *Aurora 7* was in a good re-entry attitude, and he consequently selected the auxiliary damping control mode. He kept an eye on both the rate indicator and the window at this time, as he was beginning to see the re-entry glow.

"It was actually a beautiful re-entry despite all my worries," Carpenter later recorded in the astronaut book, *We Seven*. "The ride most of the way down was perfectly smooth, and we headed in at a good angle. When I glanced out the window I could see an orange doughnut of fiery particles stretching out like a wake behind me. These were tiny bits of the ablative heat shield which had melted off and were carrying some of the intense heat away with them. Everything was normal."[15]

He could also see some flaming pieces detaching from the spacecraft. At one stage he even watched as a long rectangular restraint strap broke off and rapidly disappeared in the distance. Despite not being in touch with the Cape, he continued to record his impressions as *Aurora 7* plowed through the ever-thickening atmosphere.

"I noticed one unexpected thing during the heat pulse. I was looking for the orange glow and noticed instead a light green glow that seemed to be coming from the cylindrical section of the spacecraft. It made me feel that the trim angle was not right and that some of the surface of the recovery compartment might be overheating. However, the fact that the rates were oscillating evenly strengthened my conviction that the spacecraft was at a good trim angle. The green glow was brighter than the orange glow around the window. The period of peak acceleration was much longer than I had expected. I noticed that I had to breathe a little more forcefully in order to say normal sentences."[16]

Carpenter would later say that he was never frightened during this dangerous phase of his flight – just interested.

"The oscillations were building up now and I could feel them and hear them going 'bang, whump, bang, whump,' as the capsule swung from side to side. They were welcome, because they meant that an aerodynamic force would be exerted against the capsule and help keep it on an even keel on the way down."[17]

At around 120,000 feet the g-forces began to taper off as *Aurora 7*, its speed now reduced to about 600 miles an hour, continued to head for an ocean splashdown.

"But the swaying built up rapidly, and I used the very last of my fuel trying to control it. I was concerned that the capsule might topple over completely and start coming down topside first. If this happened, the drogue chute could get badly fouled up if it came popping out during this wild swinging, or it might snap the capsule around so violently that the chute would be badly damaged during deployment."[18]

At around 70,000 feet, the oscillations became quite bad, and *Aurora 7* started swaying through what Carpenter later described as "a huge arc of about 270 degrees – almost full circle." As he said in his post-flight pilot's report, he decided to take matters into his own hands by manually deploying the drogue chute a little early.

"I switched the drogue parachute fuse switch on at about 45,000 feet. At about 40,0000 feet, spacecraft oscillations were increasing. At about 25,000 feet, I deployed the drogue parachute manually when the oscillations became severe. I could see the drogue chute pulsing and vibrating more than I had expected. It was visible against a cloudy sky."[19]

At around 15,000 feet, with the drogue chute successfully deployed, Carpenter switched the main parachute fuse switch on. While waiting for the parachute to deploy, he decided to intervene.

"At about 9,500 feet, I manually activated the main parachute deployment switch without waiting for automatic deployment. It came out and was reefed for a little while. I could see the parachute working as the material was stretched taut and then as it undulated after the peak load. The parachute disreefed and it was beautiful. I could see no damage whatsoever, and the rate of descent was right on 30 feet per second. Convinced that the main parachute was good, I selected the automatic position on the landing bag switch, and the bag went out immediately. I went through the post-entry and 10,000 foot checklists and got everything pretty well taken care of."[20]

During his final descent at 04 53 13, a transmission from Gus Grissom came through, but it became obvious that Grissom was transmitting blind, hoping Carpenter could hear him as he reported that the spacecraft was going to land 200 miles long, that the Air Rescue people would be notified and they should be with him in about an hour. Carpenter acknowledged, but Grissom did not hear his transmission.

The MA-7 mission was rapidly coming to an end, with a splashdown just moments away. Carpenter sat back as far as he could in his couch and waited for the impact with the water.

REFERENCES

1. M. Scott Carpenter, "Pilot's Flight Report," from *Results of the Second U.S. Manned Orbital Space Flight, May 24, 1962*, NASA SP-6, MSC, Houston, TX, 1962
2. *Ibid*
3. Scott Carpenter and Kris Stoever, *For Spacious Skies: The Uncommon Journey of a Mercury Astronaut*, Harcourt, Orlando, FL, 2002
4. M. Scott Carpenter, "Pilot's Flight Report," from *Results of the Second U.S. Manned Orbital Space Flight, May 24, 1962*, NASA SP-6, MSC, Houston, TX, 1962
5. Scott Carpenter and Kris Stoever, *For Spacious Skies: The Uncommon Journey of a Mercury Astronaut*, Harcourt, Orlando, FL, 2002
6. M. Scott Carpenter, "Pilot's Flight Report," from *Results of the Second U.S. Manned Orbital Space Flight, May 24, 1962*, NASA SP-6, MSC, Houston, TX, 1962
7. Gene Kranz, *Failure Is Not an Option*, Simon and Schuster, New York, NY, 2000
8. M. Scott Carpenter, "Pilot's Flight Report," from *Results of the Second U.S. Manned Orbital Space Flight, May 24, 1962*, NASA SP-6, MSC, Houston, TX, 1962
9. Scott Carpenter and Kris Stoever, *For Spacious Skies: The Uncommon Journey of a Mercury Astronaut*, Harcourt, Orlando, FL, 2002
10. M. Scott Carpenter, "Pilot's Flight Report," from *Results of the Second U.S. Manned Orbital Space Flight, May 24, 1962*, NASA SP-6, MSC, Houston, TX, 1962
11. *Ibid*
12. Scott Carpenter, *Life* magazine article, "I got let in on the great secret," issue 8 June 1962, pg. 30
13. M. Scott Carpenter, L. Gordon Cooper, Jr., John H. Glenn, Jr., Virgil I. Grissom, Walter M. Schirra, Jr., Alan B. Shepard, Jr., Donald K. Slayton, *We Seven*, Simon and Schuster, Inc., New York, NY, 1962, pg. 458–459
14. M. Scott Carpenter, "Pilot's Flight Report," from *Results of the Second U.S. Manned Orbital Space Flight, May 24, 1962*, NASA SP-6, MSC, Houston, TX, 1962
15. M. Scott Carpenter, L. Gordon Cooper, Jr., John H. Glenn, Jr., Virgil I. Grissom, Walter M. Schirra, Jr., Alan B. Shepard, Jr., Donald K. Slayton, *We Seven*, Simon and Schuster, Inc., New York, NY, 1962, pg. 459
16. M. Scott Carpenter, "Pilot's Flight Report," from *Results of the Second U.S. Manned Orbital Space Flight, May 24, 1962*, NASA SP-6, MSC, Houston, TX, 1962
17. M. Scott Carpenter, L. Gordon Cooper, Jr., John H. Glenn, Jr., Virgil I. Grissom, Walter M. Schirra, Jr., Alan B. Shepard, Jr., Donald K. Slayton, *We Seven*, Simon and Schuster, Inc., New York, NY, 1962, pg. 460
18. *Ibid*
19. M. Scott Carpenter, "Pilot's Flight Report," from *Results of the Second U.S. Manned Orbital Space Flight, May 24, 1962*, NASA SP-6, MSC, Houston, TX, 1962
20. *Ibid*

6

Walter Cronkite: "We may have lost an astronaut"

"The landing was much less severe than I had expected," Carpenter later recounted in his official pilot's report on the MA-7 mission. "It was more noticeable by the noise than by the g-load, and I thought I had a heat-shield recontact problem of some kind. I was somewhat dismayed to see water splashed on the face of the tape recorder box immediately after impact. My fears that there might be a leak in the spacecraft appeared to be confirmed by the fact that the spacecraft did not immediately right itself.

"The spacecraft listed halfway between pitch down and yaw left. I got the proper items disconnected and waited for the spacecraft to right itself. However, the list angle did not appreciably change. I knew that I was way beyond my intended landing point, because I had heard earlier the Cape CapCom transmitting blind that there would be about an hour for recovery. I decided to get out at that time and started to egress from the spacecraft."[1]

SHIPS ON STANDBY

The aircraft carrier *Intrepid* (CVS-11) had been directed to proceed to an area approximately 200 miles east of Turks Island, Bahamas, where it was planned Carpenter would splash down if the mission ran for the scheduled three orbits. Other potential impact areas were 350 miles east of Bermuda (for one orbit) and 325 miles south of Bermuda (for two orbits). *Intrepid* departed Norfolk for NASA recovery area No. 9 on 5 May, and recovery rehearsals utilizing dummy Mercury capsules were carried out while steaming out to that station. The three-day launch delay (19 to 22 May) provided *Intrepid*'s crew with three days of R&R at St. Thomas, in the Virgin Islands. On 22 May the space shot was postponed once again and *Intrepid* carried out additional recovery operations while steaming to Roosevelt Roads Naval Station, Puerto Rico. At 11:32 a.m. on Thursday, 24 May, near *Intrepid*'s designated recovery area, all capsule recovery stations were manned.

© Springer International Publishing Switzerland 2016
C. Burgess, *Aurora 7*, Springer Praxis Books, DOI 10.1007/978-3-319-20439-0_6

USS *Intrepid* at sea, circa 1962. (Photo: U.S. Navy)

Rear Admiral Earl R. Eastwold (center of photo, pens in hand) was commander of the sea-borne recovery forces ready to pluck Scott Carpenter from the ocean. (Photo: U.S. Navy)

Soon after, word was received that Carpenter had landed several hundred miles from *Intrepid* after 4 hours, 53 minutes and 47 seconds of flight. In fact, at the time the closest surface recovery vessel was some 230 miles away. The U.S. Coast Guard in the Virgin Islands reported that the spacecraft had come down off Anegada Island, 19 degrees and 29 minutes north latitude and 64 degrees and half a minute west longitude.

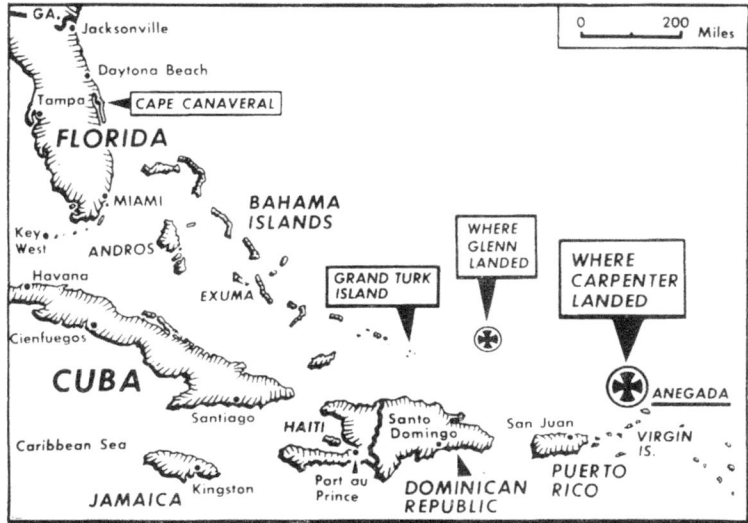

Newspaper sketch showing respective landing areas of Glenn and Carpenter. (Photo: UPI)

As Carpenter prepared to exit *Aurora 7* and wait in the safety of his life raft, the landing bag beneath the capsule began to fill – as designed – with water to act as a sea anchor. This helped to mostly right the capsule, although it still tilted to one side. However, the Mercury spacecraft was not designed to be a good boat; even the best sailor would soon become nauseous inside the stuffy cabin, and Carpenter knew he had around an hour to wait before anyone reached him. And of course the capsule was heavy and would sink quickly if a lot of water somehow managed to flood the cabin. He took off his helmet and removed the right-hand side of the instrument panel in order to make his exit. This opened up a narrow egress through the nose of the capsule. He then methodically squeezed and squirmed his way up past where the landing parachutes had previously resided.

It was far from an easy exit, but Carpenter felt it was a far better option than blowing the side hatch, with the probability of losing the spacecraft altogether. He had no desire to go through the same experience as Gus Grissom, whose *Liberty Bell 7* had sunk to the bottom of the Atlantic when the hatch blew unexpectedly and the capsule rapidly filled with sea water. However, he ought to have recalled the problem Grissom had of salt water seeping into his spacesuit, for in his haste to egress through the top, he forgot that the approved procedure called for him first to deploy the protective neck dam and seal the suit inlet hose valve.

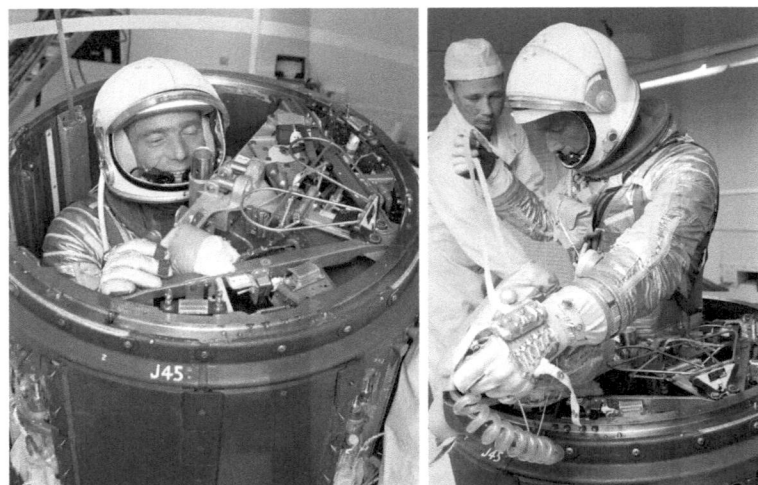

Prior to his space flight, Carpenter practiced difficult egress techniques through the top of a mock-up Mercury capsule. (Photo: NASA)

"Egress is a tough job," he later explained. "The space is tight, and the small pressure bulkhead stuck slightly. I easily pushed out the canister, and I had the life raft and the camera with me. I disconnected the hose after I had the canister nearly out. I forgot to seal the suit and deploy the neck dam. I think one of the reasons was that it was *so* hot. After landing I read 105 degrees on the cabin temperature gauge. I felt much hotter in orbit than after landing; and although it was humid, I still felt fine."[2]

Carpenter opened the hatch on the narrow top end of his spacecraft, carefully perched himself on the nose, and placed the camera on top of the recovery compartment. He then lowered himself into the water, pulling the life raft out after him, and dropped it into the water, where it inflated automatically as he hung onto the side of the spacecraft.

After he had clambered onto the life raft, Carpenter realized that it was upside down. He slid into the water once again, righted the raft, and climbed aboard. After securing the raft to *Aurora 7* to keep them together, he turned on the SARAH (Search and Rescue and Homing) beacon which would allow recovery planes to home in on his position. He realized his neck dam was still stowed, and after several fatiguing attempts he was able to deploy it some 30 minutes later. The neck dam was a safety device that provided a rubberized barrier against any further sea water seeping into his pressure suit, which already contained an undetermined, but moderate, quantity of water.

Completely alone on the water, Carpenter stretched out on the raft, relaxed, and said a brief prayer of thanks. Shortly after seeing a patch of Sargasso weed, he smiled widely when he noticed what he later described as a curious, 18-inch-long black fish that wanted nothing more than to visit. "Tame as a chicken, he was just floating in the water so close that I could have reached out and grabbed him. But I didn't because it might have hurt him, and at the time he was my only friend."[3]

Fortunately the sea was quite calm, with an occasional swell, so there was little to do but contemplate his flight and wait for the recovery aircraft to sight him. "The time on the ocean was very pleasant. I drank a lot of water from my survival kit while I was in the raft, but as far as temperature was concerned I was comfortable."[4]

The astronaut sat for a long time pondering the events of the day. As he later wrote in the Mercury book, *We Seven*, "I have never felt better or happier in my life. I felt like a million dollars. I had gotten a little water in my suit when I jumped into the ocean to right the raft, but this was not unwelcome because it kept me from getting too hot inside the suit."[5]

RECOVERY UNDER WAY

At Naval Air Station Roosevelt Roads in eastern Puerto Rico, two 12-man crews from Patrol Squadron 18 (VP-18) were preparing to take off in the squadron's Lockheed P2V-7 Neptune airplanes. As recounted by Lt. Gerald W. McDonald, their mission "was to pin-point Carpenter's location when *Aurora 7* splashed down in the Atlantic. The scheduled landing point, about 75 miles north of San Juan, was at the center of a 200-mile ellipse running from northwest to southeast along the flight path of the returning capsule. I was the pilot in command of the plane assigned to await the recovery about 50 miles from the northwest end; Lt. Jimmy Hickman's P2V would wait the same distance from the southeast end. We were at our posts about 90 minutes before the expected time of the splashdown.

"While waiting for Carpenter's arrival, our crews settled into a familiar routine, keeping the coffeepots going and checking out our equipment, chief among it a specially installed SARAH system and our standard APS-20 radar. SARAH ... had been developed by the British for rescuing downed airmen. It was a small radio transmitter, about the size of a deck of cards, for Project Mercury flights placed outside the astronaut's capsule but inside the re-entry heat shield, while a highly sensitive receiving system had been installed in the P2Vs assigned to recovery missions. APS-20 radar had been designed to seek very small targets on the ocean's surface, specifically submarine snorkels and periscopes. If SARAH failed, this radar would be used to search for the capsule, the astronaut in the raft, or, in the worst case, debris."[6]

It had been a worrying time for many people that day. From splashdown at 12:41 to 1:16 p.m. (EST) there was no word from Carpenter, although NASA knew that he had survived the flight because they had been monitoring his heartbeat throughout the fiery re-entry and they knew the spacecraft was on the ocean because the homing beacon had activated. The world's public, however, were being told through the media that the astronaut was missing. A stony-faced CBS veteran reporter Walter Cronkite was commentating live on Carpenter's flight from Cape Canaveral, and he kept up a running monologue on efforts to contact Carpenter.

Veteran news broadcaster Walter Cronkite. (Photo: CBS)

"We have a very – even more disturbing report, it seems to this reporter here, from NASA space authorities," he said to the TV camera, with concern etched across his face. "They say they did not pick up any radar blips from the descending spacecraft. It almost beggars description as to what that could mean. Whether they should have picked up radar blips of the spacecraft as far as two hundred miles away is something that experts will have to answer, and we're trying to get that answer for you. It seems to this reporter that they should have. If the *Aurora 7* spacecraft came safely through its atmospheric re-entry, it would seem that even two hundred miles would not be too far to pick up a radar signal .... So there's not even been radar contact with the *Aurora 7* since the last contact with Scott Carpenter by voice, which was that when he announced his g forces building for the re-entry into the atmosphere."[7]

At one point a little later, he gravely told the viewing audience, "While thousands watch and pray, certainly here at Cape Canaveral, the silence is almost intolerable." Then, to a suddenly stunned nation, he stated in a voice filled with emotion, "We may have ... lost an astronaut."[8]

In the event of the death of Scott Carpenter during his mission, Vice President Lyndon Johnson had prepared a statement that he would read out once the loss of the astronaut had been confirmed. It was one official statement that he hoped he would not have to use in a broadcast to the nation.

Lieutenant Commander Malcolm Scott Carpenter's name will go down in that part of history that records man's noblest efforts to reach beyond himself; his name is first on the list of Americans who went into space and did not return.

As time passes, the names of other astronauts will be added to this roll of honor. For we are going on to conquer the unknown realms around us and harness its power for the use of God's creatures on this planet Earth. This is as Scott Carpenter would have wished it. We will not disgrace his honor or his memory by turning back now.

I know I speak for the nation in its sincere but inadequate efforts to console Scott Carpenter's wonderful family and his brave associates who will carry on after him. I hope it gives them some small comfort to know that we will never forget him, and that he did not die in vain. The flight of *Aurora 7* was a necessary and important step into space that we will be able to stand on as we reach for the stars.[9]

As history records, it was Jimmy Hickman's VP-18 crew that made the initial sighting of *Aurora 7* and Scott Carpenter after nearly three-quarters of an hour had passed without the nation knowing whether he had completed a safe splashdown. Lt. Robert Goldner, co-pilot on the P2V-7 – which was basically the same type of aircraft that Carpenter had flown in Korea – was the first person to spot the floating capsule after Aviation Technician 3 (AT3) Dan Van Brocklin had picked up an electronic signal emanating from the spacecraft.[10]

A Lockheed P2V-7 Neptune similar to the one from which the crew first spotted *Aurora 7* in the water. (Photo: U.S. Navy)

FROGMEN IN THE WATER

Meanwhile an Air Force Air Rescue Squadron (ARS) SC-54 containing two pararescue frogmen had taken off from NAS Roosevelt Roads in Puerto Rico as soon as it appeared Carpenter was about to splash down. The frogmen on board were Staff Sergeant Ray McClure and Airman First Class John Reitsch (both 41st ARS). The aircraft was under the command of Capt. J. K. Eastman of the 54th Air Rescue Squadron and his co-pilot was First Lt. J. E. Powell.

After locating the spacecraft, the SC-54 roared over Carpenter. Knowing that an SA-16 triphibian capable of landing on water was on the way, the pararescuers plotted their jump pattern. Heading upwind at about 1,000 feet the lumbering SC-54 dropped a spotting chute directly over the capsule. On a second pass Capt. Eastman checked the number of seconds it had taken them from the time he flew over the spotter chute and the time he flew over the capsule again. It was about nine seconds. Eastman then made a third pass heading upwind on a direct line between the spotter chute and the capsule.[11]

At 1:48 p.m., with Heitsch leading, the two men were then able to make a precise jump from 1,200 feet carrying bags filled with 135 pounds of gear (rafts, knives, diving equipment, tools, medicine and so forth). For McClure, it was his 139th operational jump; for Heitsch, it was his first. Both hit the sea about 100 feet from the spacecraft. As their feet touched the water they pulled safety clips which released their main parachute and swam underwater to *Aurora 7*.

By this time other aircraft had begun arriving on the scene and Carpenter was distracted, watching the airplanes circling overhead. "The first one I saw was a P2V. I took out the signaling mirror from my survival kit. Since it was hazy, I had some difficulty in aiming the mirror, which is done by centering the small bright spot produced by the Sun in the center of the mirror. However, I knew the planes had spotted me because they kept circling the area. Another aid to the planes in locating me was the dye marker which was automatically ejected by the spacecraft. There must have been a stream of dye in the water 10 miles long. Soon there were a lot of airplanes around, but I just sat there minding my own business."[12]

As he was observing the airplanes circling his position, Carpenter did not see the men jump from the SC-54 or hit the water with their life rafts. Heitsch quickly reached the astronaut, swimming up behind him. Carpenter heard splashing noises and turned in surprise.

As Heitsch recalled that moment, "He looked at me with a big look of surprise on his face and said, 'How did you get here?' I told him who I was and he welcomed me." His partner, McClure, swam up soon after.[13]

"More aircraft kept circling over us," Carpenter later reported. "From time to time, one would drop a smoke bomb marker. A 20-man life raft was dropped, but the parachute failed to open and it hit the water with a tremendous impact. Attached to the raft was another package containing the Stullken collar, a flotation device much like a life preserver which can be wrapped around the spacecraft to keep it floating. It also hit with a terrific force which, as we learned later, broke one of the CO_2 bottles used to inflate the collar. The divers started out to get the collar and it took them some time to bring it back. They finally got back, wrapped the collar around the spacecraft, and inflated it."[14] Carpenter later said the capsule was listing badly and probably would have sunk without the gear, as water was gradually seeping in through the small pressure bulkhead.

The pararescuers inflated their own life rafts and climbed into them. "Then we paddled over and sat and chatted with the astronaut," said McClure. While the three bobbed on the waves, McClure said the astronaut calmly opened a water container he had with him, and like a genial host offered each of the rescuers a drink, which they accepted. "Then he pulled out a bar of concentrated food and began munching on it. He offered us some but we thanked him and turned it down."[15]

Although McClure and Heitsch spent about 40 minutes on the water with Carpenter, they were under orders not to ask questions about his flight and had to restrict their conversation to generalities. In fact Carpenter mostly led the conversation, asking the two enlisted men about their families, their work, and how often they had made jumps; he was also under instructions not to comment in any way on his space flight in order to keep it fresh in his mind.

One of the airplanes still circling overhead was the Air Rescue Grumman SA-16 Albatross amphibian that had established visual contact with the spacecraft 39 minutes after it splashed down, but the pilot, First Lt. Bruce H. Driscoll, did not land even though the water was calm. He'd heard that Carpenter was with the pararescue team and from all accounts was in good condition, and could be seen casually waving at them. The decision was made to stick to the planned routine of taking Carpenter to USS *Intrepid* for the initial debriefing, after which he would be flown to Grand Turk Island. If the SA-16 had picked up the astronaut they would have had to take him over to Puerto Rico, from where the aircraft had taken off.[16]

INTREPID'S HELICOPTERS TO THE RESCUE

Paramount in the minds of the NASA representatives on board USS *Intrepid* was the quick and safe recovery of the astronaut. However the great distance required a change of strategy. Plans were therefore altered and it was decided to use a turbo-powered Sikorsky HSS-2 helicopter, holder of the world helicopter speed record, to pluck him from the sea.

Not more than ten minutes after *Aurora 7* was located, two HSS-2 Sea King helicopters from *Intrepid*'s own Helicopter Antisubmarine Squadron 3 (HS-3) were airborne carrying NASA medical, diving, and photo crews. HS-3 Commanding Officer, Navy Capt. John Merrit ("Wondy") Wondergem piloted the prime recovery helicopter, assisted by Lt. Cdr. Billie C. Young and Lt. (Jnr. Grade) William J. Shufelt, Jr. They had originally been scheduled to fly NASA photographers to document the recovery operations. When the two HSS-2s arrived on the scene they saw Carpenter calmly sitting in a large life raft next to the capsule.

Carpenter and one of the pararescue frogmen floating in their rafts alongside *Aurora 7*. (Photo: NASA)

A pararescue frogman waves as the helicopter with photographers hovers overhead. (Photo: NASA)

Capt. Wondergem hovered while his co-pilot supervised the pick-up. The second HSS-2 remained in the area to assist and discharged two NASA divers who aided the paramedics in the recovery until everyone was hauled aboard the helicopters.

As Carpenter recalled, "When the HSS-2 helicopter appeared, it made a beautiful approach. One of the divers helped me put on the sling, and I picked up my camera which I had previously placed in the recovery compartment. I motioned to the helicopter pilot to take up the slack in the line and I let go of the spacecraft expecting to be lifted up. Instead, I went down! The helicopter must have settled slightly, because I am sure that there was a moment when nobody saw anything of me but a hand holding a camera clear of the water. A moment later, however, I began to rise. It was a lift of some 50 to 60 feet."[17]

Capt. Wondergem remembered Carpenter clutching a camera as Lt. Shufelt assisted him to board the helicopter. When Shufelt commented on this to the astronaut, he said, "I've got so much invested in this camera I wouldn't want to lose it. I went through a lot of effort to get the film." As it turned out, about half of the film had been exposed to water.

Once he was seated, Carpenter pulled the uncomfortably tight rubber collar from around his neck to allow some air to circulate over his body. He then borrowed a pocket knife, took off his left boot, and to everyone's surprise cut a hole in the left foot of his rubber pressure-suit sock, after which he stuck that leg out of the helicopter window to let some seawater and sweat drain from it into the ocean. He then lay down on the floor of the helicopter, raised his other leg, and let more water transfer through the suit's relief tube and out the window.

Carpenter is winched aboard the HSS-2 helicopter. (Photo: NASA)

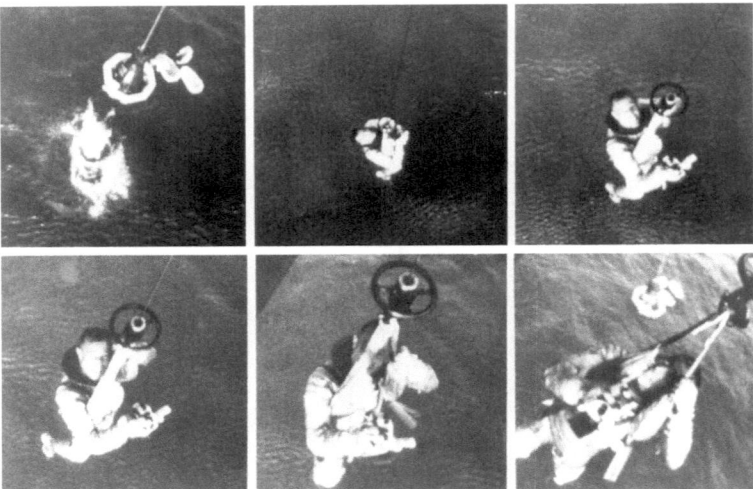

Stills from a film taken of the astronaut recovery from aboard the helicopter

A frogman ensures the collar is firmly tied to the bobbing spacecraft. (Photo: NASA)

Dr. Richard A. Rink, a lieutenant colonel in the U.S. Army, who flew in Wondergem's Navy helicopter for the pickup, said that the astronaut's greatest concern as he climbed into the helicopter had been that his suit was half filled with water. According to

Dr. Rink, the sea water in Carpenter's space suit had turned green when it became mixed with some dye that he carried to make himself a more visible target for the search planes.

An obviously exhilarated astronaut is transported by helicopter to the carrier *Intrepid*. (Photo: NASA)

When Dr. Rink had a chance to check Carpenter out during the long ride to the carrier, the astronaut was in good spirits. "He was the most exhilarated person I've ever seen. He was just bursting to tell someone about his trip, but it was part of my job to see he did not discuss it until he was debriefed and I steered away from any question which would lead him into discussing the flight."[18]

During the long ride to the waiting carrier Carpenter joked with the crew and paced about. Occasionally he would stretch and shout out, "Wow!" As Dr. Rink later reported, Carpenter "finally relaxed as one normally would after an extended mental and physical exercise."

The water-soaked astronaut had been plucked from the water at 3:30 p.m., and it took the HSS-2 an hour and 25 minutes to reach *Intrepid*, where it set down gently. The para-rescue team were retrieved by the second HSS-2.

Upon reaching *Intrepid*, some 9 hours and 10 minutes after being launched at 7:45 a.m., Carpenter was elated. "When the helicopter landed aboard the carrier, I was in good shape. Although I had already had a long day, I was not excessively tired and I was looking forward to describing my experiences to those at the debriefing site."[19]

The soaked astronaut is welcomed aboard USS *Intrepid*. (Photos: NASA)

Carpenter is escorted to a cabin for his preliminary medical check-up and an initial report on his flight. (Photo: NASA)

Meanwhile, back at the Cape Canaveral beach house, an anxious Rene Carpenter was waiting on news of her husband.

"The wait for news of him during that long period of silence after re-entry was a difficult one for all of us inside the house. But as the time passed and we could not know exactly what had happened to him, I was never really worried for his life. No matter what it was, even if the heat came through the capsule, I felt sure he would be safe. I have such faith in his physical ability to withstand anything that the worst I could think was that he might be dazed or semiconscious and that a breath of fresh air would revive him. At no time did I think he hadn't made it …. But I plumped pillows, aimlessly straightened the coffee table and waited with everyone else."[20]

Then, along with the rest of the nation, came an announcement from NASA's Shorty Powers in the Mercury Control Center that brought with it a flood of relief. "We now have an unconfirmed report from downrange of a visual sighting by a P2V aircraft in the landing zone. We are working to reconfirm that P2V report."

Soon after came confirmation. "This is Mercury Control. We have just received a report through our recovery operations branch that an aircraft in the landing area has sighted the spacecraft, and has sighted a life raft with a gentleman by the name of Carpenter riding in it."

For Rene Carpenter the relief was almost palpable. She had been worried sick about the lack of information and was deeply concerned for her husband's safety. "Then a familiar combination of sounds shocked me to attention. P2V, I heard. A P2V had picked up the sound of a SARAH beacon, and I just knew it must be the capsule. It was marvelously fitting that Scott's beloved P2V, the first plane he'd handled operationally and commanded in a squadron, should be the plane to spot him now. The room grew brighter and when that word was followed by the announcement of the actual sighting of 'a gentleman named Carpenter' the smiles came and then a brilliant burst of inane jokes which sent us into relieved and prolonged laughter."[21]

Broadcaster Walter Cronkite was also a relieved man, although he said that NASA could have done a lot more to relieve the public's anxiety by simply keeping the media informed, allowing them to report on the astronaut's well-being.

"They knew that Carpenter was alive; we did not. They had telemetry; heartbeat, respiration and so forth, and they were getting those signals. He was alive – he got through it, and they didn't tell us for 43 minutes. We were all put in the position of assuming that he was dead, and I'm wondering how we're going to tell this story and when they were going to let us know … and they never did."[22]

Immediately after the safe recovery of Carpenter had been reported to President Kennedy in Washington, he issued a statement saying, "The American people will be gratified by the successful orbital flight of Lieutenant Commander Malcolm Scott Carpenter and his subsequent rescue. The skill and initiative of those who participated in the rescue of Commander Carpenter, coupled with Commander Carpenter's courage, is heart-warming to us all."

SAFELY ON BOARD

After Carpenter had been welcomed aboard *Intrepid*, and as he prepared below decks to give his initial debriefing and undergo a more extensive medical check-up, the president called and talked with him by radio-telephone from the White House. President Kennedy offered his personal congratulations, and later authorized NASA to award its Distinguished Service Medal to the nation's newest hero. Carpenter would reveal it was not a very good telephone line, but he was able to make out the president's words.

JFK:	"Hello … hello. Hello, Scott. Come in … are we talking to Scott Carpenter?"
Carpenter:	"Mr. President, I'm … I hear you, sir."
JFK:	"Oh, well I wanted to tell you we're relieved, and very proud of your trip. I'm glad that you got picked up in good shape. And we want to tell you that we are all for you, and send you the very best luck, to you and your wife."
Carpenter:	"My apologies for not having aimed a little bit better on re-entry, as I said to people on the ship."
JFK:	"Oh, fine, good. Well, we want to congratulate you and I look forward to seeing you in Washington sometime soon."
Carpenter:	"I look forward to that, sir."
JFK:	"Very good. Good luck now, Scott. Goodbye."
Carpenter:	"Thank you very much, sir."

Scott Carpenter receives his phone called from a relieved President Kennedy. (Photos: NASA)

The earlier anxiety of *Intrepid*'s crew about the astronaut's welfare and whereabouts had quickly faded with the news that he was safe and had been retrieved by helicopter, and was replaced by a prideful exuberance that America's newest space traveler was on board their ship.

"Everybody was glad to see him," recalled retired Rear Adm. J. Lloyd ("Doc") Abbot, the ship's skipper at the time. "We were very happy to have him back. We had dinner in my cabin. The steward had brought steaks, and someone said, 'You had a hard day, Scott, you better pick a big one.' Scott looked at me and said, 'Can I have two?' I said, 'You can have them all!' It was a very, very wonderful experience."[23]

Writing later on the subject of wonderful experiences, Carpenter reflected back on his thoughts while bobbing around in his life raft waiting to be picked up. "I felt that space was so fascinating and that a flight through it was so thrilling and so overwhelming that I

His dramas now behind him, Carpenter prepares to remove his wet space suit following a safe recovery from the Atlantic. (Photo: NASA via AP Wirephoto)

only wished I could get up the next morning and go through the whole thing all over again. I wanted to be weightless again, and see the sunsets and sunrises, and watch the stars drop through the luminous layer, and learn to master that machine a little better so I could stay up longer. There's no doubt about it, space is a fabulous frontier, and we are going to solve some of its secrets and bring back many of its riches in our lifetime. I would not miss that for anything."[24]

RETRIEVING *AURORA 7*

USS *Farragut* (DLG-6), initially located about 90 nautical miles southwest of the calculated landing position, was the first ship to reach the floating capsule. The destroyer had not been a part of the actual recovery team, but was on a short operational cruise to San Juan, Puerto Rico and St. Thomas, Virgin Islands for missile firings and shore bombardment exercises. The cruise was unexpectedly climaxed by an exciting race to *Aurora 7*'s overshoot splashdown area, as recorded by an unknown correspondent in the ship's Mediterranean 1962–63 cruise book, *Looking Aft*.

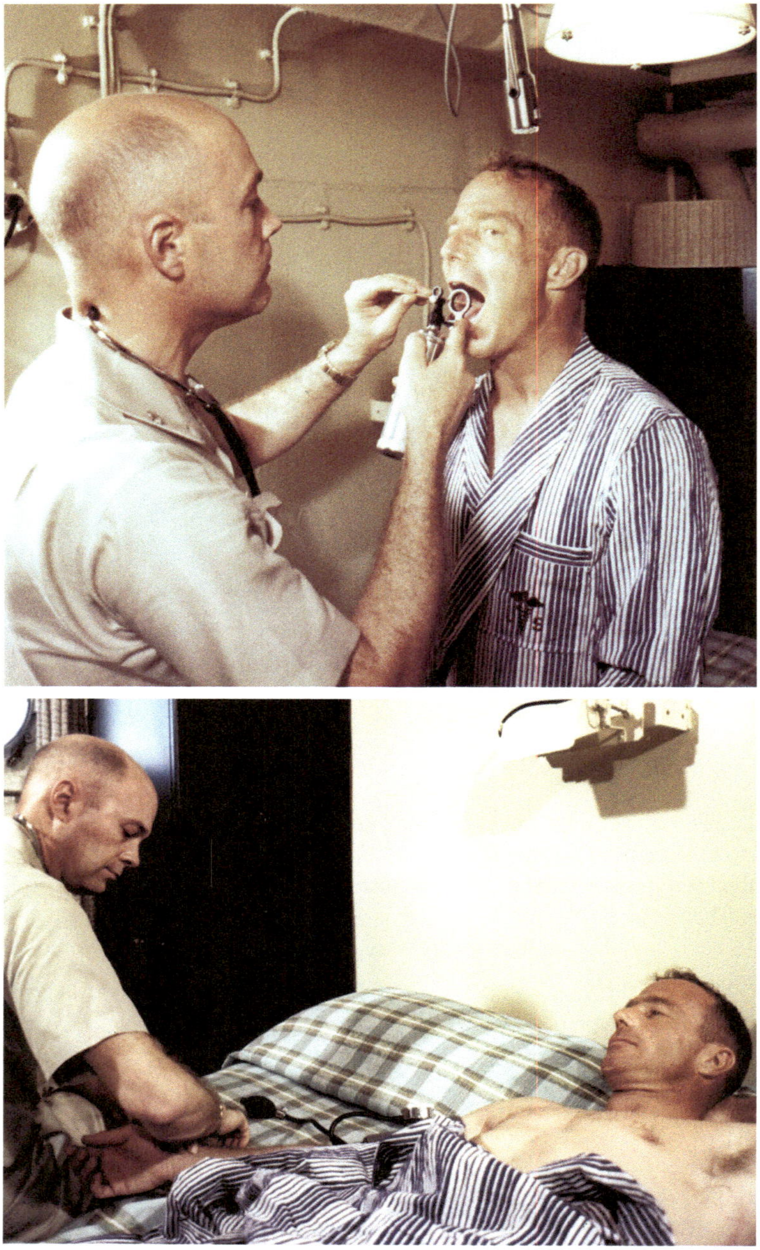

Comfortably clad in a dressing gown, Carpenter undergoes an initial medical examination. (Photos: NASA)

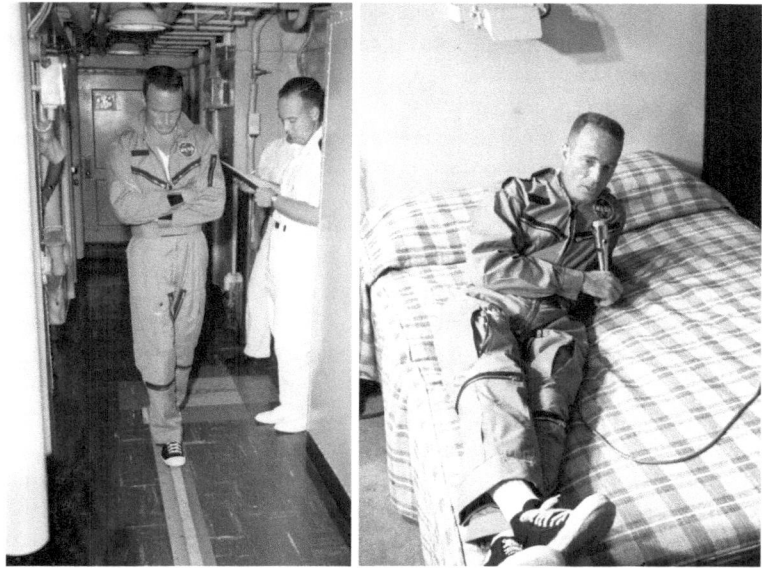

After a check of his balance was completed, Carpenter recorded his first impressions of the flight. (Photo: NASA)

USS *Farragut*. (Photo: U.S. Navy)

Underway that historical morning from San Juan, Puerto Rico, to conduct a shore bombardment exercise and then to proceed home to Mayport, it was in the role of interested Navy men and proud citizens that we listened to the news account of Carpenter's apparently flawless Mercury space shot.

But in the space age, the line separating success from failure is measured in microseconds. Suddenly we received the word: "Aurora 7 overshot the recovery force; exact whereabouts unknown." As apprehension mounted throughout the world for the astronaut's safety, orders reached Farragut to break off from our assigned mission and take part in the search and recovery.

Despite the fact that Farragut was not a member of the designated recovery group, her crew knew exactly what to do. The repair gang improvised a spanning bar for the boat davits which could lift "Aurora 7" aboard, should it be necessary. The Engineering Department brought the steam plant up to full power, while topside the crew joined in clearing the fantail for helicopter operations.

At maximum speed, the ship turned towards the probable splash area, hours away, and all hands strained to be the first to spot the missing capsule – so eager that the hours were thought of as minutes, the miles expressed as yards.

Too late – by 20 minutes! Not even one of the Navy's fastest ships can outrun a jet helicopter, which lifted LCDR Carpenter into the air and took him to safety aboard a carrier.

Farragut was the first ship on the scene, recovering all the paramedic gear and the astronaut's personal equipment and checking the flotation attachments of the capsule, until a specially rigged recovery ship arrived in the area.[25]

The *Farragut* crew maintained close watch until the spacecraft could be retrieved using special equipment aboard another ship that was due to arrive in about six hours. During that await, *Farragut* launched her motor whale boats to enable sailors to check the

This photo of a listing *Aurora 7* with USS *Farragut* in the background was snapped by an unknown crew member aboard a motor whale boat as they waited for the spacecraft to be retrieved by the destroyer USS *John R. Pierce*. (Photo: Farragut Association)

integrity of the flotation attachments and to recover the paramedic gear and Carpenter's personal equipment.

In June 1959, after completing five years at engineering school in Louisville, Kentucky, Leigh Bartlow received his Ensign commission in the U.S. Navy. He was later assigned to USS *John R. Pierce* (DD-753), a *Sumner*-class destroyer, which played a vital role in the recovery of the *Aurora 7* spacecraft. In May 1962, he was a 25-year-old lieutenant-junior grade, and recalled the activities surrounding that historic event.

"A year prior to our deployment to the primary Mercury recovery area north of Puerto Rico, my primary billet was as First Lieutenant, responsible for all of the deck force – about 60 officers and men – and associated equipment. As such, I was directly involved in the strengthening of our forward boat davit by the addition of two thick vertical welded steel plates. This davit was one of two davits on the port side that held a motor whaleboat we called the captain's gig, and allowed it to be swung out over the side of the ship and be put in the water, and recovered. The davit needed to be stronger in order to recover a Mercury spacecraft – in the unlikely event that such was ever needed. My deck crew had trained in the manual recovery procedure – ropes and all – under the tutelage of my senior BMs [Boatswain's mates]. Shortly afterwards, I left the deck department and was assigned as Ship's Navigator, the position I held during the Carpenter flight.

"The recovery force consisted of the carrier *Intrepid* and two DDs [destroyers]. We were the downrange recovery DD in the primary recovery area. The other DD was uprange of the carrier, with the three ships spread out along the 119 degree orbital track in the landing area. The carrier was the landing target for the spacecraft. We had broadcast radio pool coverage reporters (one of whom was a ham radio operator, too) on board, and they had strung wire antennas topside as necessary to communicate back to the U.S. At the expected landing time, we were steaming at 15 knots – on a course of about 030 degrees – through our designated position, at right angles to the landing track. Captain Lorensen had previously ordered full engine room capabilities – and the snipes had all four boilers, full superheat and all the rest, on line, and we were ready....

"As all may recall, Scott Carpenter overshot the landing point – in fact the entire landing area – as a result of firing his retrorockets with the spacecraft pointing high of the proper angle, and indeed landed some 250 miles downrange of the carrier – creating a real cliff hanger.

"The Atlantic Ocean land-based system for long-range sonar triangulation – SOSUS – was used to very quickly locate Carpenter. In a few minutes the SOSUS position report came by radio from the Cape over one of the bridge speakers. I plotted the position and realized that not only were we the closest recovery ship, we were still nearly six hours away, even at maximum speed. I immediately recommended to the skipper a course of 119 degrees. We came right with full rudder, and flank speed was ordered on the engine order telegraph, with all the turns we could muster! Many crewmembers will remember that ride. We kept our fingers crossed that the spacecraft wouldn't sink before we got there.

"Six hours and two hundred miles later we made a night rendezvous with the freighter that was standing by, and successfully completed the only DD recovery of a manned spacecraft, and the only night-time recovery by anyone. As the officer most familiar with these recovery procedures, I left the bridge and went to the portside recovery area to provide direction, advice and assistance as needed, standing under the glare of the lights."

USS *John R. Pierce*. (Photo: U.S. Navy)

Aurora 7 and the flotation collar are raised onto USS John R. Pierce. Note the damage to the landing bag. (Photos: NASA)

A "shepherd's crook" was used to attach a lifting line to the gently bobbing spacecraft, which was then hoisted aboard the destroyer. However, as the spacecraft was brought onto the ship it was noted that the landing bag had been damaged quite extensively and all of its straps were broken, probably caused by wave action while the spacecraft was supported by the flotation collar prior to recovery.

As Bartlow pointed out, "We also recovered the life raft used by Carpenter prior to his being plucked from the ocean by choppers from *Intrepid*."

After it was retrieved from the ocean and the spacecraft's explosively activated side hatch had been unbolted, a considerable amount of sea water was found sloshing inside

Aurora 7. This was believed to have penetrated through the small pressure bulkhead when Carpenter passed through the recovery compartment into the raft. "We did have some concern about *Aurora* sinking," Bartlow told the author. "I guess we should not have worried, since the collar system did its job. But there was about 60 gallons of water in the spacecraft when we got it aboard."

Bartlow wrapped up his account as follows, "About midnight we steamed alongside the carrier and high-lined spacecraft contractor personnel aboard. They inspected the spacecraft, now sitting on the main deck, port side – where our motor whaleboat usually

Aurora 7 being unloaded from USS *John R. Pierce*. (Photo: NASA)

resided – and drained most of the water out of it. The next day, after dropping the spacecraft and contractor personnel off at Roosevelt Roads, we headed back to Norfolk. That turned out to be my last trip on the *Pierce* – I left active duty after almost three years on her. This was pretty heady stuff for a fellow from the mountains of north-eastern Pennsylvania."[26]

The spacecraft is carefully loaded onto a flatbed truck. (NASA film footage still)

Aurora 7 on a pier at Naval Station Roosevelt Roads, Puerto Rico. From the pier it was moved to the airfield to be loaded aboard an Air Force transport plane. The aircraft in the background were Navy P2V-7 Neptunes from VP-18, which was involved in the recovery operation. (Photo: Gerald McDonald)

An interesting side note is that during the recovery operation an antenna from the spacecraft broke off and was kept by the crew of *John R. Pierce*. The machine shop made several rings for the crew from the antenna which had the words "Aurora 7" imprinted on the front side.

The spacecraft is positioned in front of the transport plane's nose ramp, ready to be flown from Puerto Rico directly to Cape Canaveral. (Photo: Gerald McDonald)

While *Aurora 7* sat on the pier at NAS Roosevelt Roads, hundreds of service personnel gathered around for a look at the history-making spacecraft. It was eventually moved from the pier to the airfield, where it was loaded onto an Air Force transport aircraft for the less dramatic return to Cape Canaveral.

GRAND TURK ISLAND

Meanwhile, the astronaut had boarded an aircraft late that Thursday night for the journey to the 12-square-mile Grand Turk Island in the Bahamas, where he would undergo a far more searching medical examination and debriefing.

Prior to the astronaut departing USS *Intrepid* the skipper, Capt. Lloyd Abbot, made a small souvenir presentation to the nation's newest space hero. (Photo: NASA)

Dressed in a blue flight suit, Carpenter prepares to leave *Intrepid* for Grand Turk Island. (Photo NASA)

A final photo opportunity as Carpenter thanks *Intrepid*'s skipper, Lloyd Abbot. (Photo: NASA)

On the flight across to Grand Turn Island, Carpenter was relaxed and in high spirits. Ahead of him were two days of medical and psychological examinations, technical and engineering reviews of the flight, and a chance to unwind before facing a daunting number of official events, presentations, and press conferences.

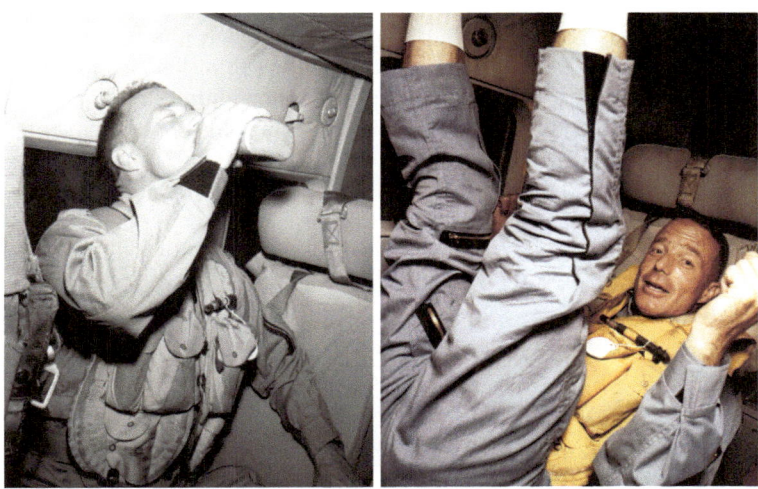

Relaxed and in good spirits, the astronaut is flown over to Grand Bahama Island. (Photos: NASA)

Although Carpenter was under instructions not to discuss his space flight until it had been properly recorded, he took the time to talk with members of his accompanying crew during the flight. (Photo: NASA)

Waiting to greet him on the island was John Glenn, who had been fielding questions from assembled members of the press. Before the Navy plane touched down, he was asked how the flight had gone. "It turned into a pretty good evening," he responded.

Glenn was then asked if he had harbored any doubts during the flight. "I don't think there was anyone on Cape Canaveral who did not have some doubts sometime during the day," he mused. "I know I did."[27]

While preparing to deplane in his blue coveralls just after midnight, some 10 hours and 39 minutes after his drama-filled splashdown, Carpenter was met with a warm embrace from an emotional Glenn plus a warm handshake from Wally Schirra. Carpenter then shook hands with Geoffrey Guy, British Administrator on the island, who said, "It's nice to have you back with us." Col. John Powers, NASA's Public Affairs Officer, then escorted the astronauts through a milling crowd of press reporters and photographers to an automobile that was to take them to the small medical center on the grounds of the island's U.S. Air Force auxiliary base. Carpenter smiled broadly as he and his colleagues walked briskly to the vehicle, but offered no comment apart from saying cheerily, "Hello everybody; I'll tell you all about it tomorrow." They were then driven directly to the medical facility where they were greeted by flight surgeons Howard Minners and Bill Douglas.

John Glenn and Scott Carpenter give each other a joyful hug on Carpenter's arrival at Grand Turk Island. (Photo: NASA)

A brief medical examination was undertaken inside the island's base hospital about an hour after the astronaut's arrival. Dr. Minners had already talked to the physicians aboard *Intrepid*, who had pronounced Carpenter to be in excellent shape, so he was not expecting anything different.[28]

An internist member of the debriefing team stated that Carpenter "entered the dispensary with the air and the greeting of a man who had been away from his friends for a long time. He was alert, desiring to tell of his adventure and seemed very fit … his appearance and movements suggested strength and excellent neuromuscular coordination."

In fact, Carpenter was still filled with understandable enthusiasm over his flight. "Boy, the sunrises and sunsets," he told the physicians, engineers and space flight technicians as he recounted his entire day, before he forgot any of it. "They are more beautiful than anything I've seen on this Earth." Asked whether he had any anxious moments, Carpenter volunteered that he had "a few moments of anxiety near the retrofire, over whether I was going to have enough fuel." Glenn and Schirra sat up talking with Carpenter until exhaustion set in and he retired to bed at around 3:30 a.m.

Two other Mercury astronauts, Gus Grissom and Gordon Cooper, would join the others on Grand Turk. Deke Slayton was returning from Australia, where he had served as a CapCom. He was being flown directly to Florida and would meet up with his colleagues at Patrick Air Force Base.

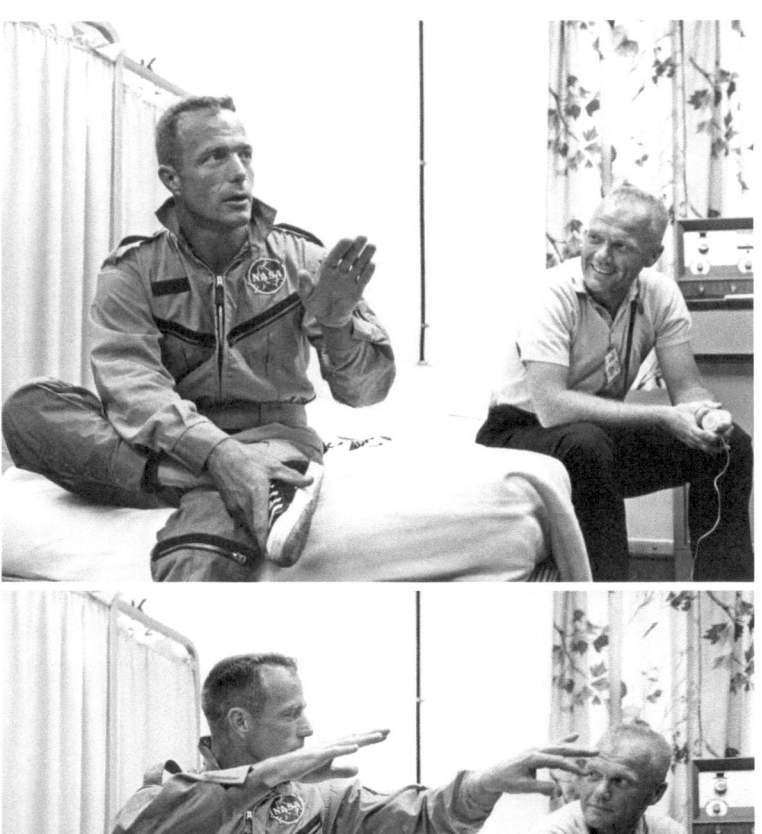

Scott Carpenter relives his orbital flight for fellow astronaut John Glenn (Photos: NASA)

A FAMILY REJOICES

On Thursday evening, buoyed by the safe return of her husband, Rene Carpenter emerged from the beach house with her children to speak at a motel press conference in Cocoa Beach. On stage with her were the four children: Robyn Jay, Kristen Elaine, Candace Noxon, and Mark Scott Carpenter.

In her opening address Rene revealed that her husband had phoned her from the space-craft prior to lifting off. She added that she looked forward to a reunion with him on Sunday, and they intended to take a trip to Colorado. Rene further remarked that the day of the flight was one filled with tension and "did not match the day three years ago when we first learned Scott had been selected as an astronaut."

In a voice filled with emotion, she said, "I want to say that the effort involved in one of these missions is such that at the end we often feel emotionally drained and we tend to fall back on comfortable phrases like happy, proud and thrilled and we feel so much more.

"Truly these men and this project belong to you and the country as nothing else in my memory. I want to say that not once in these past three years would I have had Scott do any less than he did today. For those who serve, the rewards are very great."

Rene Carpenter and the four children on stage at a press conference in Cocoa Beach. (Photo: NASA)

Rene said that she went through the harrowing hours of the flight without tears or prayers, even when he overshot the landing area and was out of contact for more than 40 minutes. "I gave a little pause when the newsreader said contact … had been lost, but I had been thoroughly checked out and was familiar with all the procedures."

Twelve-year-old Scotty displayed the only small doubt in the family when he said he thought he wanted to be an astronaut but changed his mind during the minutes between retrorocket firing and the moment his father was sighted floating in the ocean. His brother Jay, 10, was asked if he still wanted to be an astronaut when he grew up. "Well …" he said, "in a way."

Rene mentioned that their two daughters were too young to be space-conscious, and they had played on the beach during the launch, while the two boys took more notice.

When asked how she had managed to keep in seclusion until that evening, Rene smiled as she responded, "It was very easy – dedicated families and the U.S. Navy." She was not about to reveal anything to do with the *Life* magazine beach house. "The Navy brought me down and friends loaned me a house, stocked it with food, and left me alone." When pressed for details she simply replied that she would not divulge any names, saying that they were dear friends and that "the privacy afforded me was wonderful."[29]

Meanwhile, Florence ("Toye") Carpenter told reporters at a press conference held in the ballroom of the University of Colorado campus that she had watched the launch and progress of her son's space flight on television. She had been with two neighbors and Will Fowler, a former classmate of Scott's at the university, at her green and white trailer home, located in a neat court in northeastern Boulder.

Toye mentioned that every weekday morning she went to work at the Boulder Community Hospital, about a mile away, where she was in charge of the medical records department. On Sunday mornings she faithfully attended Episcopalian services. Still frail and thin from her long fight with tuberculosis, her eyes shone as she talked of her Bud – although she referred to him as Scott, knowing that in interviews it would only confuse matters if she talked about him using the nickname by which she always called him. Toye was asked by one reporter if his becoming an astronaut had any effect on their close mother-and-son relationship.

"No, I don't think so," she pondered. "If anything has changed, it has been that I have come to have more respect for the qualities which have enabled him to be what he is. He is a very unassuming man. He is very thoughtful of other people. He is very kind to me. I wasn't really worried. Scott is not a worrier, nor am I. It is a senseless waste of energy to worry about things I cannot control. I can't afford to do that. My hope is always up."

The televised launch was certainly a dramatic sight, she said, but her son's recovery from the sea had obviously been the high point of the day for her. She added that she had never doubted that he would be recovered because "he was conditioned for it, he was trained for it and he was capable."

Asked whether she thought he might be the first person on the Moon, she replied without hesitation, "I'd like to have him wait a few days until I catch my breath!"[30]

At their mountain lodge home in Palmer Lake, 75 miles south of Boulder, Scott's 60-year-old retired chemist father Marion had viewed the mission along with his second wife Edythe, whom he had married in 1944. He told a reporter, "That was the tensest five hours I've ever spent." He added, "The most tense time, of course, was when we learned the capsule had overshot the mark. Naturally, we were worried. Before that, there was the worry over the vehicle's possible burning up during re-entry. But I had faith in our technology. I'm a very proud man, and I'm indeed glad it's over. Now that it's over, I can say that I hope he is chosen to be a member of the first three-man team to orbit the Earth."[31]

PRESS CONFERENCES AND PRESENTATIONS

After Carpenter awoke on Grand Turk Island at 9:15 on Friday morning, a comprehensive medical examination was carried out by the same group of specialists who had examined him a week before his space flight.

Aside from moderate fatigue, based upon the long hours of work and a few hours of sleep, the astronaut remained in excellent health throughout the aeromedical debriefing period. He then underwent an engineering debriefing. In spite of growing concerns in the media that Carpenter might have been overly distracted during his flight, or too tired to land correctly, the examiners were so pleased with the results of their checks that they said NASA should consider extending manned space flights beyond three orbits.

The morning after his arrival at Grand Turk Island, Carpenter and Glenn were interviewed by the waiting media. (Photo: NASA)

After two days of aeromedical and engineering debriefing and relaxation, including three pleasant hours of scuba diving, Carpenter was flown back to the Cape by military plane on Sunday, 27 May.

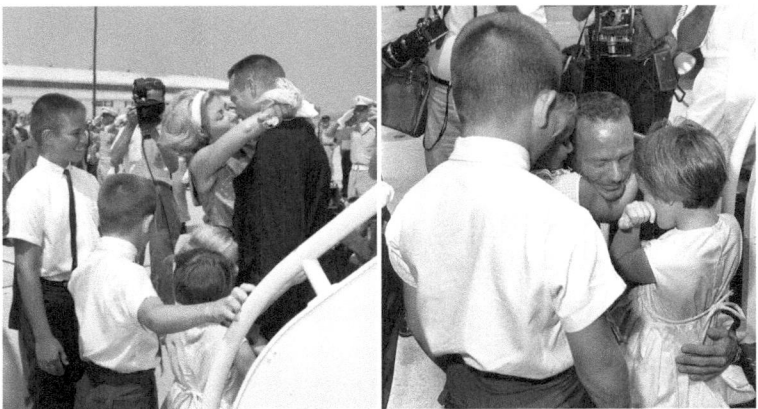

On arrival at Patrick Air Force Base, Carpenter is greeted by his family, receiving a kiss from his wife Rene and hugs from his children. (Photos: NASA)

Flanked by his mother Toye and wife Rene, Carpenter gives a short speech to the crowd at Patrick Air Force Base. (Photo: NASA)

On arrival at Patrick Air Force Base he was greeted by his wife Rene and their children, as well as some two thousand friends and well-wishers, who gave him a hero's welcome. After giving Rene a long kiss, he grabbed and hugged each of his children. He told the cheering crowd that his three-orbit flight was "the supreme experience of my life," and then added, "I am glad to be back."

The triumphant motorcade from Patrick AFB through Cocoa Beach (Photo: UPI)

The Carpenter family had meanwhile been joined by NASA Administrator James E. Webb for a triumphant motorcade through the town of Cocoa Beach, waving at happy crowds lining the streets, waving flags and shouting their congratulations as the automobiles moved on to Cape Canaveral.

The party went directly to Hangar S at the end of the Cape where, in a brief ceremony, both Carpenter and Walt Williams, operations director of Project Mercury, were presented with NASA's Distinguished Service Medal by James Webb. Said Carpenter in part, "I accept this humbly on behalf of all the people on whom these flights depend and without whom they would not be possible …. I am uneasy in this acclaim. I am aware that it is due and merited by at least a thousand people here. This is the hardest working group of people I have ever seen."

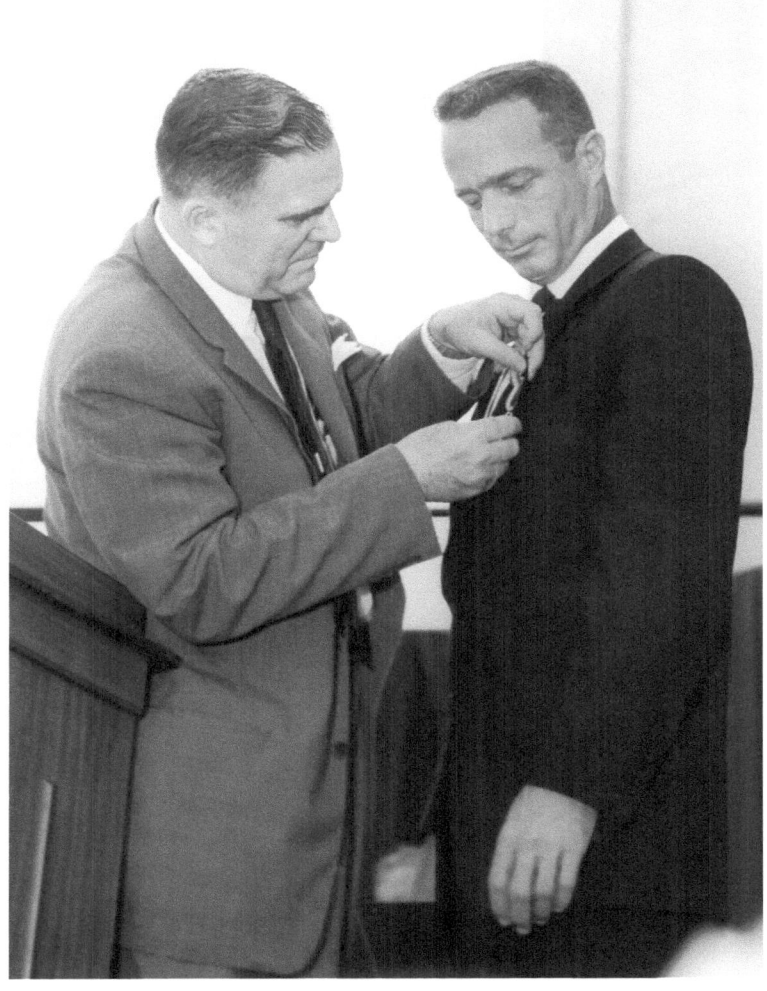

Scott Carpenter is presented with NASA's Distinguished Service Medal by NASA Administrator James Webb. (Photo: NASA)

Following the presentation, the group made their way over to where *Aurora 7* had been temporarily positioned for the event. Carpenter spent several minutes giving his family a close-up inspection of the spacecraft.

Following the medal presentation, Carpenter was able to show his family members the interior of the historic *Aurora 7* spacecraft. (Photos: NASA)

The next scheduled event was a press conference at the Cape press site, where Carpenter responded to questions about his three-orbit journey into space and back, at one stage saying he was "ready to do it again."

Obviously, many of the questions from reporters related to the unexpected overshoot and the time that he was "lost" at sea. Carpenter explained that all the fuel for his manual control system was expended after the retrorocket firing. He pointed out that he had sufficient fuel aboard, but "I managed it improperly." He recalled for the press conference that at the very beginning of the first orbit he had found that the automatic stabilization control system was operating improperly. Since this was to be largely a manually controlled flight, he was not particularly concerned at the time. "There are so many things to do. Everything you see is such an awe-inspiring sight that you don't have time to linger on things that aren't of immediate importance."

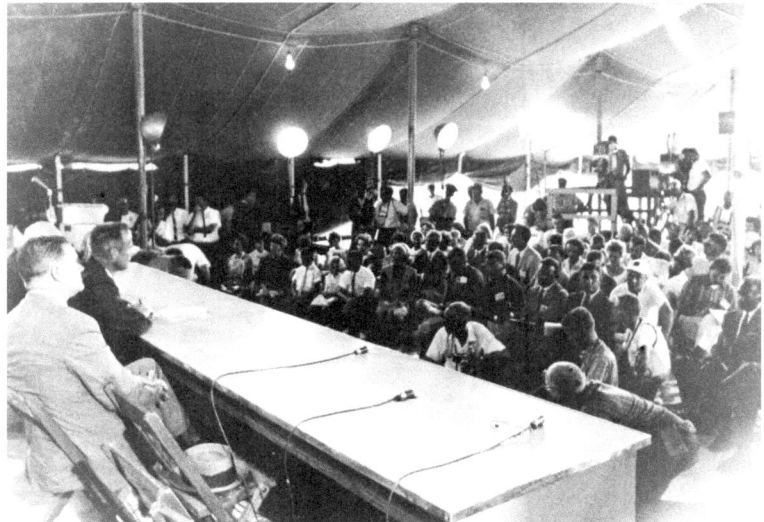

Scott Carpenter addresses the media and responds to questions about his flight. (Photo: NASA)

The Carpenter family seated in the front row of the press conference. (Photo: NASA)

Carpenter said he went on with his flight plan and tried to accomplish all of the objectives. "On the third orbit, south of Hawaii, by count at the debriefing, there were six things going on all at one time, all of which were important. I understand the report came from Hawaii that I was a tired and confused astronaut. If my opinion is worth anything to you this is not true. I will admit to being preoccupied. It was a very busy time."

Asked about some of his impressions from the flight, Carpenter responded that the most arresting sight was the sunset or sunrise. "It's a beauty beyond description; the pictures will be available later and you can judge for yourself." He said that the flight was "the supreme experience of my lifetime."

The following day, Scott was greeted by his father, Marion Carpenter, at Denver Airport, Colorado. (AP Photo/WPS)

At the time, cosmonaut Yuri Gagarin was visiting Osaka, Japan, as part of a worldwide propaganda tour and he expressed his good wishes for Carpenter. He also told a cheering crowd of 5,000 people that "the Soviet Union dreams of the day cosmonauts of various countries will be riding in the same space ship." However, Gagarin introduced a touch of competitiveness into an interview with Japanese newsmen when he said that Carpenter's flight, "cannot be termed a total success" because he missed his landing by a considerable margin. "There were no such failures with either myself or Gherman Titov," he said. "We landed in our designated areas."

What Gagarin was not allowed to reveal was that he had almost died during the re-entry phase when the descent module on his Vostok spacecraft remained attached to its jettisoned equipment module by a steel cable that was supposed to have separated, causing the vehicle to tumble wildly through the ferocious heat of re-entry. Fortunately for Gagarin, the cable finally burned though, the two modules flew apart, and the descent module stabilized just in time with the protective heat shield facing into the worst of the heat. But it had been a very close thing.

Gagarin also did not mention that both he and Titov had actually parachuted out of their spacecraft above the ground due to the lack of retrorockets to slow the spacecraft just above the ground. The Vostok capsule, on its own parachute, was designed to land hard, without the cosmonauts on board.

To emphasize the Cold War competitiveness, Gagarin then told the reporters that, "It is still unknown whether Carpenter's flight is a fact. However, if it is true, it did not indicate any progress" – over John's earlier three-orbit flight.[32]

PILOT'S SUMMARY

In giving his pilot's report on his flight, Carpenter summed up his feelings on his own performance, and that of *Aurora 7*.

> Overall, I believe the MA-7 flight can be considered another successful step on the road to the development of a useful and reliable manned spacecraft system. The good performance of most of the spacecraft systems gave me confidence in the vehicle itself, while the spectacular novelty of the view from space challenged me to make the most of my opportunity, and lured me into an unwise expenditure of fuel early in the flight. As a result, it became necessary to go to extended drifting flight, and I was able to demonstrate that there was no problem associated with prolonged drifting flight, a procedure we shall have to make use of on the longer duration Mercury flights.
>
> I was able to detect and overcome the one significant systems malfunction that might have affected the flight: the malfunction of the pitch horizon scanner circuit. I understand that many [people] were concerned while waiting without word from me during re-entry and after landing. However, from my position, there was no major cause for concern. The spacecraft was stable during the critical portions of re-entry and the parachute worked perfectly.
>
> For me, this flight was a wonderful experience, and I anxiously await another space mission.[33]

That second flight would never happen for Scott Carpenter.

REFERENCES

1. M. Scott Carpenter, "Pilot's Report," from *Results of the Second U.S. Manned Orbital Space Flight*, May 24, 1962, pg. 74, NASA SP-6, Washington D.C., 1962
2. *Ibid*
3. Scott Carpenter, LIFE magazine article, "I got let in on the great secret," issue 8 June 1962, pg. 35
4. *Ibid*
5. Carpenter, S., Cooper, Jr. L, Glenn, Jr., J., Grissom, V., Schirra, Jr., W., Shepard, Jr., A., and Slayton, D., *We Seven*, Simon and Schuster Inc., New York, NY, 1962, pg. 463
6. Gerald W. McDonald, article, "The Recovery of 'Aurora'," *American Heritage* magazine, Vol. 52, No. 2
7. Walter Cronkite, live CBS television report, 24 May 1962
8. *Ibid*

9. Civil Archives Branch of the National Archives: the Reading File of the Executive Secretary, National Aeronautics and Space Council (RG 220), March 1961–June 1973; Fldr: Chron. File, May 1962

10. JAX Air News, unaccredited article, "Lt. Goldner of VP-18 Was First to Sight Astronaut," U.S. NAS Jacksonville, Florida, Vol. 20, No. 9, 31 May 1962, pg. 1

11. *Salt Lake Tribune* newspaper, article "Three Times 'Round – Safe!" Salt Lake City, Utah, issue 25 May 1962, pg. 8

12. M. Scott Carpenter, "Pilot's Report," from *Results of the Second U.S. Manned Orbital Space Flight*, May 24, 1962, pg. 74, NASA SP-6, Washington D.C., 1962

13. Edward V. McCarthy, *Milwaukee Sentinel* newspaper article "Astro Hosts 'Tea Party' on Ocean for his Rescuers," issue 25 May 1962, pg. 12

14. M. Scott Carpenter, "Pilot's Report," from *Results of the Second U.S. Manned Orbital Space Flight*, May 24, 1962, pg. 74, NASA SP-6, Washington D.C., 1962

15. Edward V. McCarthy, *Milwaukee Sentinel* newspaper article "Astro Hosts 'Tea Party' on Ocean for his Rescuers," issue 25 May 1962, pg. 12

16. *Airescue Information Newsletter*, "41st Jumpers Save Aurora 7, Aid Astronaut," Vol. 1, No. 1, 15 June 1962

17. M. Scott Carpenter, "Pilot's Report," from *Results of the Second U.S. Manned Orbital Space Flight*, May 24, 1962, pg. 74, NASA SP-6, Washington D.C., 1962

18. *Ibid*

19. *Ibid*

20. Rene Carpenter, LIFE magazine article, *Scott Carpenter and his son and his wife living through 'the time that grew too long'*, issue 1 June 1962, pg. 30

21. *Ibid*

22. Michael Lennick interview with Walter Cronkite, published on YouTube, 10 October 2013. Available at *http://youtu.be/sUknCWBNFtY*

23. Stephanie Gaskell, *New York Daily News*, article, "Bid for retired NASA space shuttle touts early Intrepid astronaut rescues," issue 27 March 2010

24. Carpenter, S., Cooper, Jr. L, Glenn, Jr., J., Grissom, V., Schirra, Jr., W., Shepard, Jr., A., and Slayton, D., *We Seven*, Simon and Schuster Inc., New York, NY, 1962, pg. 465

25. USS *Farragut* (DLG-6) Mediterranean Cruise Book *Looking Aft*, 1962–63

26. Leigh Bartlow, e-mail correspondence with the author, 31 July 2014

27. *The Miami News* (Florida) newspaper, article "Two Happy Heroes Meet in the Bahamas," issue 25 May 1962, pp. 1A and 6A

28. *Ibid*

29. *Schenectady Gazette* (New York) newspaper, article, "Space Fright Spared, Dauntless Wife Says," issue 25 May 1962, pg. 11

30. *The Miami Herald* (Florida) newspaper article, "How His Parents Waited and Hoped," issue 25 May 1962, pg. 6A

31. *Schenectady Gazette* (New York) newspaper, article, "Space Fright Spared, Dauntless Wife Says," issue 25 May 1962, pg. 11

32. Spokane (Washington) *Daily Chronicle* newspaper article, "First Orbiter Pays Respect to Carpenter," issue 26 May 1962, pg. 2

33. M. Scott Carpenter, "Pilot's Report," from *Results of the Second U.S. Manned Orbital Space Flight*, May 24, 1962, pg. 74, NASA SP-6, Washington D.C., 1962.

7

From astronaut to aquanaut

Five days after he had orbited the Earth three times, a man who once believed he was failing a university course in heat transference told professors at the University of Colorado what it was like to pass through extremely high temperatures, at times bordering on around 3,000 degrees Fahrenheit.

After graduating from Boulder High in 1943, a young Scott Carpenter had enrolled at the University of Colorado, but only attended one semester before joining the Navy's V-12a program, designed to help train pilots during World War II. "That's what every 17-year-old wanted to do," Carpenter said of joining the military. He then returned to the university to study aeronautical engineering, but in 1949 rejoined the Navy and left the university still short of passing a course in heat transference. A successful completion of the subject would have enabled him to graduate, but he was one course requirement short.[1]

A BELATED DEGREE

Following his presentation at the University of Colorado on 29 May 1962, Scott Carpenter belatedly received his bachelor of science degree in aeronautical engineering. The president of the university, Quigg Newton, who presented a beaming Carpenter with his degree, had entertained Scott and Rene the night before at a guitar-playing sing-song. It was a good mood evening and the happy astronaut had even taught Newton's ukulele-playing daughters Abby and Ginna his favorite tune, "Yellow Bird." When presenting the long-overdue degree to Carpenter, Newton said, "His subsequent training as an astronaut has more than made up for his deficiency in the subject of heat transfer."[2]

On 5 June, Carpenter, Walt Williams and their respective families traveled to Washington, D.C. and New York to receive additional honors. In Washington, they were offered personal congratulations from President Kennedy, who said, "I cannot imagine better representatives of what we like to think our country stands for than the … men who have taken part in these flights."

© Springer International Publishing Switzerland 2016
C. Burgess, *Aurora 7*, Springer Praxis Books, DOI 10.1007/978-3-319-20439-0_7

University of Colorado President Quigg Newton welcomes the graduating astronaut to his Boulder home. Rene Carpenter is on the left, Mrs. Newton stands between the two men. (Photo: United Press International)

Carpenter's itinerary was supposed to have precluded a ticket-tape parade through the streets of New York, such as the one given to John Glenn a few months earlier. But to his delight he got one anyway as his motorcade made its way from Manhattan to the Waldorf Hotel, where he was the guest of honor at a civic reception. Thousands of cheering New Yorkers showered the astronaut and his wife with ticker-tape and confetti. At the luncheon he received a warm, three-minute standing ovation from the crowd of 1,500 people. Included in that number were two former presidents: Herbert Hoover and Harry Truman. In making a short speech of congratulations, Truman said, "He has done something I wouldn't do, even if he promised to give me the Moon if he could." Hoover said that all Americans were grateful for Scott Carpenter's feat, which "gave us a lift and raised the prestige of the United States." Following the luncheon, the astronaut was presented with New York's Gold Medal of Honor.

On 7 June, Carpenter returned in triumph for a second post-flight visit to his home town of Boulder. Among other planned events in those two days of celebrations, he would make a presentation to the Martin Company, manufacturers of the Titan missile which would feature in the forthcoming Gemini series of flights, and open a baseball game at the newly named Scott Carpenter Park.

On the first day, Carpenter addressed a huge crowd at Boulder High School, from which he had graduated 19 years earlier. In his address, he described his many experiences during his space flight and told the 3,016 graduates that it was entirely possible one of their number could become the first person to land on the planet Mars. He encouraged the graduating seniors not to be impatient, to be self-reliant, to be true to themselves, and

A delighted Scott Carpenter belatedly receives his engineering degree. (Photo: NASA)

above all else to continue their studies and broaden their education. "You really don't know the riches that await you," he advised the students, "but they're not going to drop into your lap. You must go get them."[3]

The next day, after visiting the Martin plant in Littleton (later Lockheed Martin Space Systems) and thanking them "for making possible the work we are doing in Project Mercury," Carpenter was driven back to Boulder, where he was honored at the Alumni Luncheon, held at the University of Colorado.

In the Fish Room of the White House, President John Kennedy greets Scott Carpenter and Walter Williams and their respective families. In lower photograph, Carpenter delivers remarks regarding his Mercury orbital flight. Left to right: Robyn Jay Carpenter; Marc Scott Carpenter; Rene Carpenter; Scott Carpenter; Candace Noxon Carpenter and Kristen Elaine Carpenter (both in front); President John F. Kennedy; Director of Operations for Project Mercury, Dr. Walter C. Williams; Elizabeth Ann Williams (in front); and Helen Manning Williams. (Photos: Abbie Rowe, White House Photographs, John F. Kennedy Presidential Library and Museum, Boston)

Flanked by former U.S. Presidents Herbert Hoover and Harry S. Truman, Walter Williams and Scott Carpenter receive New York's Gold Medal of Honor. (Photo: NASA)

Scott and Rene Carpenter wave to excited crowds lining the streets in his hometown of Boulder, Colorado. (Photo: NASA)

Scott Carpenter signs the register at a hotel after the family's arrival in his hometown of Boulder, Colorado. (Photo: United Press International)

"Scott always gave credit to the 'team,' not himself or the other astronauts when speaking to the public," according to his good friend Ed Buckbee. "He loved to answer questions of youngsters, always reminding them they could do anything they wanted to do in a country that encouraged innovation and creativity."[4]

Carpenter then made an appearance in front of a cheering crowd of around 10,000 in the Folsom Stadium on the campus of the University of Colorado. Later that day, and 14 years after failing to finish the course in heat transfer, he proudly stood before 1,821 of his fellow graduates, as well as his mother and father, at the university's commencement ceremonies in the Macky Auditorium. Dressed in traditional black gown and mortar board, he thanked the university for granting him the degree in aeronautical engineering he had received during the Scott Carpenter Day festivities on 29 May.

Addressing fellow graduates in his traditional gown and mortar board. (Photo: NASA)

That afternoon he visited the Scott Carpenter Park on 30th Street to throw out the first baseball in opening night ceremonies for Boulder's Kid Leagues. Once known as Valverdan Park – a former sewage disposal and dump site which was transformed into a park in 1957 – it had been recently renamed to honor the hometown hero. The early idea to create a Moon Mountain playground would prove to be too costly so a giant steel rocket ship was installed instead. This hallmark feature of the park remains a source of enjoyment to the present day. In his pre-game speech he admonished the youngsters there "to play according to the rules and to play against your opponent as you want them to play against you."[5]

That evening Carpenter's father drove him out to the airport, to catch a late-night flight to Houston.

A section of Scott Carpenter Park with the playground's steel rocket landmark. (Photo: City of Boulder Parks and Recreation)

GROUND CONTROL AND THE ASTRONAUT

In an interview for the JSC Oral History program, former astronaut training officer Robert Voas commented on the relationship and decision-making between ground control and an astronaut pilot, in which he compared the Mercury orbital flights of John Glenn and Scott Carpenter. He mentioned how ground control kept news of a possible and potentially fatal

Robert Voas (right) in discussion with Carpenter. (Photo: NASA)

NASA Flight Director Chris Kraft inside the Mercury Control Center. (Photo: NASA)

slip in the heat shield from Glenn during his MA-6 mission, saying it was later accepted to have been an error in judgment. On future flights astronauts would be briefed on any such major problems.

"In John's flight, I think we had another indication of the sensitivity of this ground control-astronaut or -pilot relationship in the decision that Chris Kraft made not to inform John about the potentially developing problem with the heat shield, which turned out … just to be a false alarm …. Chris sort of says that he felt John had enough to worry about to not tell him about that. But … this over-protectiveness kind of thing … the pilots want to know everything … and the more you keep them in the dark, the more threatened they are. So it's a misjudgment to do that. As it turned out, everything was fine. John was a little bit upset when he found out about it, but there wasn't any real concern there.

"The whole issue came up more in Scott Carpenter's flight because of the overshoot particularly and the issue about the fuel supply. But, again, there was this sort of issue of who was in charge and who was taking control. But that was probably the first example of a flight we had where the spacecraft might not have come back had there not been an astronaut in it, because the attitude control system, sensor system, failed, and it was necessary for Scott to take over manually and use that window … which had a [scribe] line in it for where the horizon was to be had. You could use that to make sure that you had the nose-down position that you needed in firing the retrorockets.

"It was more difficult to tell whether the nose was pointing in the right direction, that is, the yaw angle, so-called, because to do that visually through your window, you had to see the ground running below you and see that it was streaming straight out behind you. If it was going crosswise then you weren't pointing in the right direction, you needed to come [left or right]. But at night there wasn't enough configuration on the ground to make that very easy. So [Scott] had to use a combination of systems to line up the vehicle … that worked out well except for going further downrange …. I think it's an interesting thing, because from Scott's angle, I think he thought that the flight had just gone very well."[6]

In fact, Carpenter said during a 1990 interview that no one really took him to task at the time over his actions during the flight.

"No, it wasn't necessary, you know. I think there was some undercover displeasure, but I wasn't really aware of that for quite some time afterwards. And, you know, no one wants that to happen to the space program in the first place, so there's an honest effort to not make an issue of it – but it did get some noses out of joint. That is not uncommon in this business. Every flight is a compromise between conflicting purposes espoused by lots of different factions and organizations. The doctors want something, the scientists want something, the flight-test people want something else, the guy who flies wants his own, and it's a big compromise. You could just never satisfy everybody."

When asked whether Chris Kraft had ever directly said to Carpenter that he held him responsible for the splashdown overshoot, the astronaut reflected for a moment.

"He never said that to me, but I've heard that, of course. There's never been any animosity on my part towards him. I recognize that sort of thing as inevitable in the conduct of things; of experimental test flights like this. I don't … I feel charitable toward him for his views, and I think that he probably – well, I don't know. I can't comment for him. I never heard that, but I have read it."[7]

In his position as a flight controller during the flight of *Aurora 7*, Gene Kranz spoke of the increasing concern in the Mercury Control Center the longer the MA-7 mission continued.

"Scotty was just a different type of astronaut from the other ones. He was more lyrical is probably the word I'd say. He was more interested in exploring the environment than piloting the Mercury capsule. But he was hired as a pilot, as opposed to exploring the environment.

"Throughout the entire flight it became very obvious that we were getting behind in the use of the attitude control propellants, to the point where by … the second orbit, we'd advised Scotty to go into periods of drifting flight. The third [orbit] we told him basically to shut down and move into drifting flight, and as he arrived over Hawaii for the orbit he had his cameras out trying to take pictures of the 'snowflakes' … he was attempting to identify their source, as opposed to getting ready for retrofire, to the point where he was not in the orientation he needed for retrofire and he was behind in his checklist."[8]

Others were far more scathing in their assessment of Carpenter's performance during the MA-7 mission. In particular, the previously mentioned and influential Flight Director Chris Kraft, who devoted an entire chapter of his 2001 memoir to slamming the capabilities and actions of the Mercury astronaut. Deke Slayton recalled Kraft being "pretty incensed" about the way he felt Carpenter had handled the flight, even to the point of vowing that Carpenter would never get another flight. The quote that was reportedly echoing around the corridors of Houston post-flight was, "That son of a bitch is never gonna fly for me again."

Legendary NASA Flight Director Gene Kranz. (Photo: NASA)

In responding to a book review of Kraft's memoir, Carpenter sent a letter to the *New York Times* in April 2001, in which he vigorously defended his performance back in 1962.

To the Editor:

Regarding Henry S. F. Cooper Jr.'s review of "Flight: My Life in Mission Control" (March 11), Chris Kraft's new memoir:

John Glenn's historic flight aboard Friendship 7 took place more than 39 years ago. I was his backup pilot. Rightly or wrongly, Deke Slayton was ruled ineligible for the follow-up flight because of a minor heart condition. Rightly or wrongly, I was named to take his place.

All manner of hard feelings stemmed from these two unpopular decisions, which I did not make and over which I had no control. I took the flight, named my capsule Aurora 7 and trained hard for the six weeks I had. The flight plan, originally Deke's, called for a number of radical space maneuvers, more photography, more observation and some experiments, all of which I accomplished, in addition to returning safely to Earth with the capsule unharmed. These facts alone should be enough to vindicate the flight of Aurora 7.

But there was even more: the system failures I encountered during the flight would have resulted in loss of the capsule and total mission failure had a man not been aboard. My post-flight debriefings and reports led, in turn, to important changes in capsule design and future flight plans. These too I consider major contributions to our knowledge about spaceflight and to the successes that Wally Schirra and Gordo Cooper met with in their own subsequent flights.

Chris Kraft and I have always been at odds about my flight. Yet in his review of "Flight," Henry Cooper seems unaware of the dispute. Chris's style, he says admiringly, succeeds in "pulling no punches," as though this promises a candid and true book, when it could instead be merely vindictive and skewed. A question lingers in this rattled old septuagenarian brain: why is Chris still in the ring throwing punches?

Finally, the reviewer notes with interest Chris's early, stark realization that "things happened so fast in rocketry that an astronaut couldn't do anything" without help from the ground. A debatable proposition. In any event, Chris never acknowledged that the reverse also holds true: in space things happen so fast that only the pilot knows what to do, and even ground control can't help. Maybe that's why he is still fuming after all these years.

Scott Carpenter
Vail, Colo.[9]

In a subsequent interview, Chris Kraft would not budge from his criticism of Carpenter's performance on the MA-7 mission.

"I don't want to appear to condemn Scott Carpenter," he said. "[He] is a fine person, but nevertheless he was not what I consider to be a competent test pilot, and we wanted test pilots there because they might find themselves in a position that he indeed found himself in, and he was very fortunate that he was able to get the spacecraft down without serious trouble."[10]

Scott's response to these comments demonstrated an ongoing antagonism between the two men, fueled by certain derogatory comments made in Kraft's 2001 autobiography.

"He and I have been on opposite sides of the appraisal of my flight. He thinks it involves the failure of the man, and I think it involves the failure of the machine. And I think that there's no meeting between the two of us."[11]

A WHOLE NEW DIRECTION

"Scott Carpenter is one of Boulder's most distinguished heroes," the current mayor of that city, Matthew Appelbaum, told the author. "After Mr. Carpenter's historic space flight, the City of Boulder declared May 29, 1962 as Scott Carpenter Day, and renamed the former Valverdan Park after him. As the first human ever to penetrate both inner and outer space, holding the dual title Astronaut/Aquanaut, he embodies Boulder's innovative spirit and dedication to scientific exploration. His pioneering spirit is shared by the community and is an inspiration to many Boulderites who follow in his footsteps."[12]

Back in 1962, as indicated by Mayor Appelbaum, community leaders in his hometown of Boulder had proudly renamed Valverdan Park as Scott Carpenter Park and approved a bond to incorporate a swimming pool there. The Scott Carpenter Pool was completed in time and dedicated by Carpenter one year later, on 30 May 1963. Following his dedication speech, the hometown astronaut was picked up and ceremoniously thrown into the pool, fully clothed, by the then Mayor John Power Holloway and City Manager Robert Turner.[13]

A drenched astronaut after being ceremoniously tossed into the Scott Carpenter Pool in Boulder, Colorado. (Photo: City of Boulder Parks and Recreation)

In 1963, while awaiting news of a second space mission, Carpenter assisted in monitoring the design and development of the innovative lunar module for the Apollo program, as well as assisting with underwater training for future flight crews. He also served for a time as executive assistant to Robert Gilruth, Director of the Manned Spacecraft Center in Houston. During this period, Carpenter says he "became fascinated by the underwater work being done by the French oceanographer [Jacques] Cousteau in his Conshelf program," and saw "many parallels between that work and the work being done by the American space program." The Conshelf program was a devoted effort to create an environment in which divers could live and work on the floor of the sea. Carpenter was convinced that living conditions for human beings underwater were almost identical with zero-gravity living conditions in space.[14] But first he had to overcome a personal dread he had harbored for much of his life.

"I worked a while for Bob [Gilruth]. I didn't care too much for that, but I still kept a finger on the pulse of space flight. But it was not long after that that I decided I would like to continue with the underwater work. I was fascinated with that, and it's something I hadn't done. It also was something that – that's sort of a long story, but I was afraid of the deep ocean open water."[15]

For as long as he could recall, Carpenter had had what he called an unreasonable dread of the sea, but it took an incident that he recalled from back in his Navy days to convince him it was a fear he knew he simply must conquer.

"I flew big airplanes with a large crew out of Hawaii early in my Navy career," he says. "We were doing a survival exercise in which we had to manage ourselves in two life rafts on deep, dark, blue water. We lost overboard from the raft I was in a corner reflector, which is the most important piece of equipment you've got on a raft in a real survival situation. It is the thing the radar will pick up, and guide rescue [in] your direction. It went overboard, and I thought of trying to get it. But I was afraid of the sharks and the critters in that water, and I didn't do it. But my gunner's mate, without a second thought, jumped overboard. He was gone for a long time, but he swam down and got that corner reflector and brought it back up. And I thought, 'There is a brave man,' and it made me ashamed of myself. That was the genesis of my need to conquer my fear of the deep ocean. It's an important thing. Conquering of fear is one of life's greatest pleasures, and it can be done a lot of different places."[16]

As related by daughter Kris Stoever in Carpenter's memoir *For Spacious Skies*, his life was about to experience a profound change of direction. "Some time after the conclusion of Project Mercury, perhaps early in 1964, Deke Slayton, now coordinator of astronaut activities, sent a questionnaire around asking who wanted a Gemini flight, the two-person training missions to prepare for the lunar expeditions. Scott said no, he wasn't interested. For one thing, he'd just met his long-time hero, Jacques-Yves Cousteau."[17]

As Carpenter once observed about his wanting to team up with the legendary underwater explorer, "Curiosity is a thread that goes through all of my activity. I've been curious. I've also been frightened by the deep ocean. I wanted, number one, to learn about it; but, number two, I wanted to get rid of what I felt was an unreasoned fear of the deep water. I was also inspired by what Cousteau had done. I saw a use for NASA technology in ocean technology, and first proposed to Cousteau that I come and share technology with his program. He said, 'Well, we could use your experience, but you don't speak the right language and we can't pay you very much.' But, he said, 'if you want to share technology with the ocean, do it with your own United States Navy.' And that's how it happened."[18]

His interest piqued, he discussed the issue of working alongside Cousteau and the U.S. Navy with his NASA boss Bob Gilruth, who agreed to grant him leave to help in applying some of NASA's training and technology to the sea floor. Cousteau was equally excited about the prospect, because it meant he would have access to some excellent U.S. Navy resources.

Underwater explorer Jacques Cousteau with Scott Carpenter. (Photo: NASA)

PROJECT SEALAB

Carpenter took a leave of absence from NASA on 26 June 1964 to participate in the Navy's SEALAB project. He was scheduled to join the SEALAB I team off the coast of Bermuda around the 1st of July.

SEALAB I was an experimental underwater habitat developed by the Navy to prove the viability of saturation diving by having subjects live in isolation on the sea bed for extended periods of time. It was hoped that the knowledge gained from this, the first of three planned SEALAB expeditions, would help advance the science of deep sea diving and rescue, and contribute to the understanding of the psychological and physiological strains that humans can endure. SEALAB I was a pressurized steel habitat 12 feet in diameter and 58 feet long that was constructed from two converted floats, held in place with axles from railroad cars. It was planned to lower this prototype habitat to a depth of around 200 feet on 6 July, where it would remain submerged for over three weeks while a series of experiments and studies were performed by a team of five divers. The underwater laboratory would be pressurized to the equivalent of seven atmospheres, while a unique gas mixture of 85 percent helium, 11.5 percent nitrogen and 3.5 percent oxygen would be breathed by the inhabitants. Entrance to the laboratory would be achieved through an opening in the bottom of the habitat, with the seven atmospheres of pressure preventing any water from entering the unit.

Clearing his desk after taking temporary leave from NASA in June 1964. (Photo: NASA)

In addition to Carpenter, the SEALAB I team consisted of a submarine doctor, a chief quartermaster, a chief hospitalman, and a first class gunner's mate. Studies and tests in which he planned to be involved during his undersea tenure included such things as core drilling, the effects of corrosion, ecological studies, psychological and physiological evaluations, underwater navigation and sound propagation. One interesting study he was looking forward to involved playing taped sounds of porpoises and whales while observing and photographing the reactions of sharks and barracuda through the lab's windows. The plan was for Carpenter to join the dive team on the 13th day of the venture, then spend 12 days working with them. On completing post-dive reports and findings for the Navy Department, he was expected to return to NASA to recommence his duties at the Manned Spacecraft Center sometime in August 1964.

SEALAB I at the U.S. Naval Station, Bermuda, in 1964. (Photo: U.S. Navy)

On 16 July, Carpenter and a Navy doctor were based in Hamilton, Bermuda, returning from checking underwater lights and cameras that were intended to be used on SEALAB I, when Carpenter was involved in a serious motor scooter accident. In order to avoid colliding with an oncoming vehicle, he steered his scooter into a driveway on a gravel shoulder, but lost control, skidded wildly, and was hurled against a coral wall. He suffered a compound fracture of his lower left arm, contusions on the left knee, and the distal joint in his left big toe was crushed. He was treated at the scene and then flown to the Methodist Hospital in Houston, where surgeons tended to his injuries. They also took the opportunity during his hospitalization to break and reset a bone in his right foot which had mended improperly following the accident in his hotted-up roadster back in 1946.

The fracture of his ulna – one of the two long bones in the forearm – immediately eliminated a disappointed Carpenter from his planned participation in SEALAB I. The project nevertheless took place when the pressurized habitat, under the command of project chief Capt. George Bond, was lowered off the coast of Bermuda. Four divers took part, intending to remain at the 192-foot mark for three weeks of work and study. However, the team was obliged to surface after just 11 days due to the threat of an approaching tropical storm.

Beginning on 17 September 1964, a somewhat reluctant Carpenter resumed working as executive assistant to Robert Gilruth at MSC, albeit with his arm in a cast. It was a position he would hold for an indefinite period while convalescing from his injuries.

The following month, on Saturday morning, 31 October 1964, NASA lost its first astronaut: 34-year-old Air Force Captain Theodore Cordy ("Ted") Freeman was killed when his T-38 Talon training jet suddenly flew into a flock of snow geese over Ellington Air Force

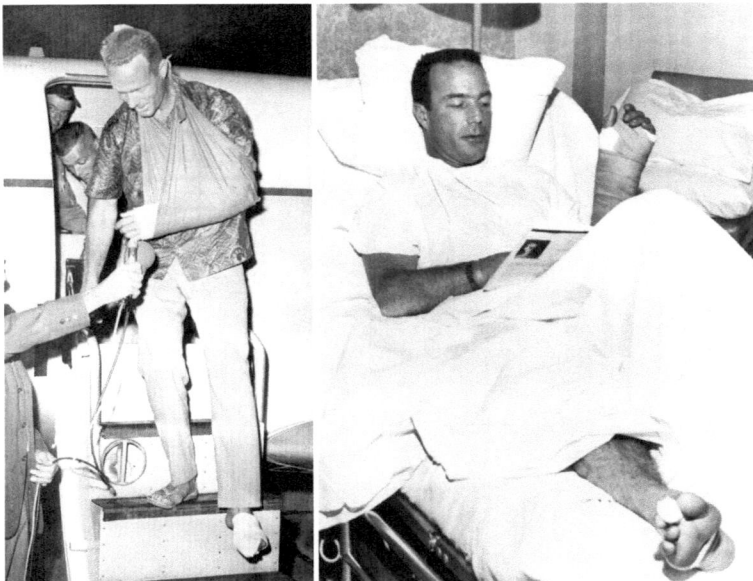

At left, Carpenter returns to Houston following his motor scooter accident. He was then hospitalized for surgery on his injuries. (Photos: United Press International)

Base on a routine training flight. One bird smashed into the Plexiglas cockpit canopy, causing the canopy to explode into fragments. These splinters were ingested into the air-breathing intakes of both engines, causing them to flame out. Freeman tried desperately to regain control over his aircraft, but ultimately had to bail out. Unfortunately he ejected too late, his parachute did not fully deploy in time, and he died instantly when he hit the ground.

Three weeks later, on 23 November, Scott Carpenter wrote and mailed a poignant letter addressed to Freeman's widow, Faith, and the Freeman's young daughter, known as Faithie, in which he summed up the tragic loss of an outstanding astronaut colleague.[19]

Dear Faith,

All of us in the course of our lives suffer the loss of dear ones. For me, this has been the loss of a son and my mother. Everlasting life for the two loved ones was a subject of concern to me during those times, and I came upon a certain realization that I sincerely hope will be of some comfort to you as well.

Ted lives on as surely as if he were physically seated by you at this instant. He lives on in Faithie and all he has contributed to her person. He lives on in your love and that of his parents and his family. He lives on in the respect and esteem of his fellow astronauts and when we get to the moon we will carry his contribution to that objective, and his name, with us just as certainly as if he were making the flight himself.

I must also say that I think you are a brave and handsome lady and I am very proud to know you.

Sincerely,
M. Scott Carpenter

Once he had mostly recovered from his injuries, and back on leave once again from NASA, he finally managed to participate as an aquanaut in the SEALAB II project. Now attached to the Navy Defense Mine Laboratory in Panama City to train 26 aquanauts, he was fortunate not to have been even more seriously injured when he was charged with driving on the wrong side of the road in Panama City, Florida, after he apparently – according to the police report – fell asleep at the wheel of his car and collided head-on with another vehicle. The two occupants of the other car were admitted to hospital, thankfully with minor injuries. Carpenter, who somehow escaped with only scratches, later appeared before a judge for a brief hearing and was released after paying a 25-dollar fine.[20]

The SEALAB II team. Carpenter kneeling in front row, second from left. (Photo: U.S. Navy)

In his capacity as an aquanaut with SEALAB II, Carpenter acted as Training Officer for the dive crew and was Officer-in-Charge of the submerged diving teams during the operation. The Navy experiment took place while the SEALAB II habitat was anchored in 205 feet of water half a mile off the coast of La Jolla, California, and linked by television to the shore. A trained bottlenose dolphin named Tuffy from the Navy's Marine Mammal Program ferried supplies down to the aquanauts. The SEALAB II habitat contained a special laboratory, a watch station, a galley, hot water showers, toilets, eleven viewing ports, and living space to support 10 aquanauts at any given time.

SEALAB II secured aboard Navy barge YC-982 (Photo: U.S. Navy)

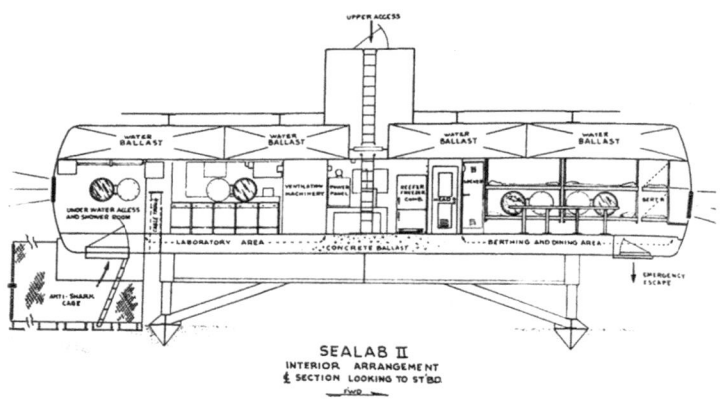

Cutaway view of the SEALAB II habitat. (Photo: U.S. Navy)

The divers, including Carpenter, entered the craft on 28 August 1965 for 45 days of tests to determine how well a person could live and work under pressure 6.5 times greater than on land. It was planned that three teams of 10 men each would live for 15 days in SEALAB II, while Carpenter would attempt an unbroken 30-day sojourn.

SEALAB II being lowered into the water. (Photo: U.S. Navy)

Scott Carpenter gives the signal to lower SEALAB II into the ocean off La Jolla, California. (Photo: U.S. Navy)

DANGERS OF THE DEEP

One unplanned but nevertheless historic event occurred on Carpenter's first day submerged, when he exchanged greetings with Gordon Cooper aboard the Gemini V spacecraft, then on its 117th orbit of a planned 120 orbits. Cooper's fellow astronaut Pete Conrad was awake, but Cooper did all the talking. It was later reported that Carpenter's voice came through in falsetto tones owing to the effect on his vocal chords of the high pressure oxygen-helium mixture of the habitat. After establishing contact, Carpenter told Cooper, "You're doing a great job. We hope you have a very pleasant re-entry shortly." Unfortunately, most of the conversation relayed by the new Mission Control Center in Houston was unintelligible.[21]

While living on the floor of the ocean, the dive teams undertook many planned activities including continuous physiological training; they tested a variety of prototype underwater gear, set up an undersea weather station, conducted extensive marine research, made exploratory dives into a nearby submarine canyon and practiced salvaging a sunken fighter jet.

Scott Carpenter working aboard SEALAB II. (Photo: U.S. Navy)

For longer excursions outside of SEALAB II, the divers did not use conventional diving systems. Instead they used a habitat-connected hose commonly called a "hookah," which supplied a diver's breathing gas from their underwater base, giving the advantage of largely unlimited endurance. Hookah equipment was used almost exclusively on SEALAB II, but the long hoses would prove to be something of an encumbrance for the divers attempting underwater activity, and there was the prevailing danger associated with a kinked hose, as Carpenter found out during one of his dives.

"All of a sudden, my supply of breathing gas was cut off – as if the pumps had failed, which I thought was the case. What do you do when you're 200 feet down, with nothing to breathe? I managed a breath-holding swim back to the SEALAB's entrance hatch. As I entered, I called to the diving supervisor to send someone out to help my 'buddy' diver, Bill Coffman, thinking he would be in trouble, too. But within seconds he popped up beside me in the hatch. He had seen my hasty retreat and, like a good buddy, had followed me home to make sure everything was all right.

"What had happened, we found, was that my hose had kinked – and after removing the kink, we finished our dive."[22]

On the 15th day, Carpenter and the nine men were swimming outside, preparing for the next shift of divers to occupy the habitat. As he was adjusting a cable on the transfer tank that would lift the men to the surface he was stung by a venomous scorpion fish.

"The scorpion fish stung me one day when I was in the water near the hatch and made the mistake of groping in the dark with my hand. The spines on the fish's dorsal fin caught me on the right index finger, and the pain was instant, a sharp and poisonous feeling. Though I never saw the fish, I knew what it was; we had noted hundreds of them – six to eighteen inches long – around the lab in recent days.

Television images of Carpenter and the other aquanauts aboard SEALAB II were broadcast live to American audiences. In the lower photo, the hand of another diver can be seen during live television pictures holding up a scorpion fish to Carpenter. Several days later, he would receive a painful sting from a larger version of that same poisonous fish. (Photos: United Press International)

"I got out of the water, undressed and went to bed immediately. The pain spread from my finger to my hand and arm and became excruciating. My nose and sinuses filled with fluid and I had to breathe completely by mouth. Bob Sonnenburg, the diver-doctor on board moved fast. He gave me antihistamine shots, cortisone and pain killers, but I hurt. There was some discussion with the surface as to whether or not I should come up right then with the first team. We decided I could stay down but would probably have to stay out of the water for five days. As it turned out, I felt somewhat better in a few hours, was much improved the next day, and was in the water the day after that. Still, it could have been a bad thing, and it would have been a great disappointment to me to have to go to the surface before my scheduled 30 days were up."[23]

A second team of nine aquanauts descended to SEALAB II after the first group surfaced. In order to help complete the 45-day experiment, they would get less oxygen than the first team. As scientists were seeking the most efficient breathing mixture for use in underwater projects, the oxygen content of the atmosphere would be reduced from 4.24 to 3.5 percent.

"Welcome to the Tiltin' Hilton" reads the sign. Due to the slope of the site at the La Jolla Canyon, the SEALAB habitat was perched at an angle on the sea bed. Scott Carpenter is in the center of the group. (Photo: U.S. Navy)

After spending 29 days and 14 hours aboard SEALAB II, Carpenter rose the 205 feet to the surface along with the second team of nine aquanauts. No one in the history of diving had ever spent as much time as the pioneering astronaut at that depth. The dive team was required to spend 33 hours in a decompression chamber aboard the staging vessel *Berkone*, with helium gas replacing nitrogen. Meanwhile, a third team of 10 men would occupy the seabed habitat for the remaining 15 days of the phased experiment.

From Texas, President Lyndon Johnson spoke to Carpenter a few hours after entering the decompression chamber. In part, the President said, "Scott, you have convinced me and all the nation that whether you are going up or down you have the skill to do the job."

Asked how he felt, Carpenter responded in a helium-fueled Donald Duck-like voice, "We are feeling very well. We entered the decompression chamber about four hours ago and everyone is happy and we're running on schedule."

The U.S. Navy SEALAB team. Scott Carpenter is fifth from left, front row. (Photo: U.S. Navy)

For his participation in the SEALAB experiment, Carpenter was awarded the Navy's Legion of Merit award.

"SEALAB was an apartment but it was very crowded," he recalled in a 1969 interview. "Ten men lived inside. We worked very hard. We slept very little." Years later, he said he actually preferred his experience on the ocean floor to his time in space. "In the overall scheme of things, it's the underdog in terms of funding and public interest. They're both very important explorations. One is much more glorious than the other. Both have tremendous potential."[24]

Scott Carpenter, astronaut and aquanaut, shows models of his two adventure vehicles to his daughters Kris and Candy. (Photo: World Book Encyclopedia Science Service Inc./Del Borer)

NASA, ever keen to put a positive spin on the achievements of its astronauts, credited Carpenter with being the first person to explore both of humanity's great remaining frontiers: the ocean and space.

After participating in the SEALAB II experiment, Carpenter returned briefly to the space program and was responsible for liaising with the Navy in underwater zero-gravity or neutral buoyancy training. But the motorcycle accident in Bermuda had caused permanent damage to his left arm. Despite undergoing a surgical procedure in 1964 (and again in 1967) he was never able to regain full mobility in that arm. As this disqualified him from a second space flight, he reluctantly resigned from NASA on 10 August 1967. "I was medically grounded. I couldn't have a Gemini or Apollo flight, even if I wanted one."[25] So he resumed work with the Navy's Deep Submergence Systems Project, serving as director of SEALAB III aquanaut operations, which focused on the development of the deep sea diving techniques required for rescue, salvage and research.

"SEALAB III was a very ambitious experiment which would have repeated much of the work done by the previous two SEALAB experiments but at the much greater depth of 600 feet," Carpenter subsequently wrote. "After many delays, equipment failures, and other major difficulties, including flooding of the habitat, and finally, the loss of Barry Cannon, one of the divers, the troublesome project was canceled."[26]

While undertaking repairs to a leak in the SEALAB III habitat off San Clemente Island, California, diver Barry Cannon died of carbon dioxide poisoning. It was later discovered that his breathing apparatus lacked baralime, the chemical used to remove carbon dioxide.

By 1968, Scott Carpenter was living in California doing underwater research while Rene and their children had moved to Washington. Their marriage had been on shaky ground for some time, and formally ended in 1972. As Rene later mused, "He's a lovely man; we just went separate ways." Rene had her own syndicated newspaper column called *A Woman, Still*, and following the divorce she had her own TV show called *Everywoman*. She was renowned for tackling difficult and often controversial subjects, such as feminism and birth control. She would marry again in 1977.

LIFE GOES ON

Scott Carpenter never did return to space, but new challenges beckoned and his explorations continued. After 25 years of distinguished service he retired from the Navy on 1 July 1969, founded and became CEO of Sea Sciences Inc., a venture capital firm which focused on the development of programs aimed at enhanced utilization of ocean resources and the improved health of the planet. He also worked closely with Jacques Cousteau and dove in most of the world's oceans, including under the ice in the Arctic.

In 1972, following his divorce from Rene, he married for a second time to Maria Roach, the daughter of famed film producer Hal Roach. They would have two children, Matthew Scott and Nicholas Andre Carpenter, who followed the family tradition by becoming a

Scott Carpenter with author Colin Burgess at a 2003 book signing in the Reuben H. Fleet Science Center, San Diego. (Photo: Francis French)

Scott Carpenter made a guest appearance at the Astronaut Hall of Fame induction ceremony in 2004. Fellow Mercury astronaut Gordon Cooper can be seen in the background at right. (Photo: NASA/KSC)

film director. Their marriage ended in 1986. Two years later Carpenter married for a third time to Barbara Curtin. They would have a son, Zachary Scott, but this marriage would also end in divorce. In 1999 he married for the final time, to Patty Barrett.

In later years he became an engineering consultant, a wasp breeder, and an author. His first book, *Inner Space*, was published in 1969. It looked at aspects of undersea exploration, including SEALAB. His first novel, entitled *The Steel Albatross* was a techno-thriller in the same vein as Tom Clancy, about a Soviet plot to place a doomsday device on the ocean floor. The follow-on novel, *Deep Flight*, was set aboard a revolutionary stealth submarine. In 2003 his long-awaited autobiography, *For Spacious Skies: The Uncommon Journey of a Mercury Astronaut*, which he wrote with his daughter Kris Stoever, was released to critical acclaim, and quickly made its way to the top of the *New York Times* best-seller list.

In 1981 Scott Carpenter was inducted into the International Space Hall of Fame at the New Mexico Museum of Space History, and in 1990 was inducted into the U.S. Astronaut Hall of Fame in Titusville, Florida. His many awards included the Navy's Legion of Merit, the Distinguished Flying Cross, the NASA Distinguished Service Medal, Navy Astronaut Wings, the University of Colorado Recognition Medal, the Collier Trophy, the New York City Gold Medal of Honor, the Elisha Kent Kane Medal, the Ustica Gold Trident, and the Boy Scouts of America Silver Buffalo. He had also been awarded seven honorary degrees.

REFERENCES

1. Mitchell Byars, *Daily Camera* newspaper (Boulder, Colorado), article "Boulder-bred astronaut Scott Carpenter – in town for park ceremony Thursday – reflects on spaceflight," issue 19 September 2012
2. Sydney *Daily Telegraph* newspaper, unaccredited article "Spaceman to give them a hot time," issue 30 May 1962
3. NASA *Space News Roundup*, Manned Spacecraft Center, Houston, TX, issue Vol. 1, No. 18, 27 June 1962, pg. 5
4. Ed Buckbee email correspondence with Colin Burgess, 24 December 2014
5. NASA *Space News Roundup*, Manned Spacecraft Center, Houston, TX, issue Vol. 1, No. 18, 27 June 1962, pg. 5
6. Robert B. Voas interview with Summer Chick Bergen for NASA JSC Oral History program, Vienna, Virginia, 19 May 2002
7. Eddie Pugh and Nigel Macknight, "Flashback Mercury 7, Part Two," from *Space Flight News*, No. 54, June 1990, pg. 24
8. Michael Lennick interview with Gene Kranz, uploaded to YouTube, 10 October 2013. Available at: *http://youtu.be/sUknCWBNFtY*
9. Joel Achenbach, *Washington Post* newspaper article, "Scott Carpenter, Mercury 7 astronaut and second American to orbit Earth, dies at 88," issue 10 October 2013
10. Michael Lennick interview with Chris Kraft, uploaded to YouTube, 10 October 2013. Available at: *http://youtu.be/sUknCWBNFtY*
11. Scott Carpenter interviewed by Roy Neal for NASA JSC Oral History program, Houston, Texas, 27 January 1999
12. Email message to Colin Burgess from Mayor Appelbaum's office, Boulder, Colorado, 23 December 2014
13. City of Boulder, Colorado website, Scott Carpenter Park and Pool History, available at *https://bouldercolorado.gov/parks-rec/history-of-scott-carpenter-park-and-pool*
14. Michael Lennick interview with Scott Carpenter, uploaded to YouTube, 10 October 2013. Available at: *http://youtu.be/sUknCWBNFtY*
15. William Harwood, CBS News article "Scott Carpenter, Mercury Astronaut, dies at 88," CBS Interactive Inc., 10 October 2013
16. Scott Carpenter interview with Michelle Kelly, JSC Oral History program, Houston, Texas, 30 March 1998
17. Scott Carpenter and Kris Stoever, *For Spacious Skies: The Uncommon Journey of a Mercury Astronaut*, Houghton Mifflin Harcourt, Boston, MA, 2003
18. Scott Carpenter interviewed by Roy Neal for NASA JSC Oral History program, Houston, Texas, 27 January 1999
19. Letter to Faith Freeman from Scott Carpenter dated 24 November 1964. From the Scott Carpenter collection, reproduced with permission.
20. Undated 1964 newspaper clippings from Sydney *Daily Telegraph*, articles "Space man out of orbit" and "Safer in space," from author's collection
21. *Sydney Morning Herald* newspaper, unaccredited article "Undersea aquanauts get space phone call," issue 30 August 1965
22. Scott Carpenter, article, "New Underwater Breathing Systems," *Popular Science* magazine, issue November 1969, pg. 72
23. Scott Carpenter, *LIFE* magazine article, "200 Feet Down, the Next U.S. Frontier," issue 15 October 1965, pg. 104
24. *The Australian* newspaper unaccredited article, "America mourns astronaut Scott Carpenter, a man from the 'right stuff' days of space travel," issue 11 October 2013
25. "Frequently Asked Questions for Scott Carpenter," from *scottcarpenter.com*
26. William Harwood, CBS News article "Scott Carpenter, Mercury Astronaut, dies at 88," CBS Interactive Inc., 10 October 2013

8

Epilogue: The man and his spacecraft

Following the ocean retrieval of *Aurora 7*, the spacecraft was shipped to Cape Canaveral where it was returned to Hangar S. A thorough visual inspection was made of the external and internal areas, and a record made of all the switch and control positions. The spacecraft was then taken to the pyrotechnic area for external disassembly and inspection. Following this, *Aurora 7* was transported to the power area for a post-flight systems check.

A de-salting wash-down, tank drainage, and flushing procedure was carried out, and deterioration safeguards were taken in general. The immediate post-flight inspection procedure included external disassembly of the heat shield and conical shingles in order to inspect the pressure bulkhead and internal skin areas. Samples of insulation were removed and stored for later analysis.

Shingles on the conical section of the spacecraft showed the usual bluish and orange tinge, while those on the upper, cylindrical section displayed the usual dark yellow-gray appearance – both caused by aerodynamic heating – but essentially the spacecraft had experienced no in-flight damage. Several shingles were slightly dented and scratched, but as in previous missions this was attributed to scrapes and bumps suffered during the recovery operation. The external surface of the heat shield had the normal, evenly charred, glass-streaked appearance, and a number of cracks were found in the ablation shield exterior, but these cracks had not compromised the safety of the mission. However, the landing bag had suffered extensive damage, and all the landing bag straps were broken, primarily due to sea action.

The worst damage was found in the spacecraft's interior. Almost the entire interior was wet from sea water that had made its way into the cabin after splashdown. About four inches of water remained in Carpenter's couch and the battery compartments. Some electrical connectors and internal spacecraft systems were heavily corroded, but all of the systems responded well to post-flight systems checks. The window was clear, although some moisture was present between the two outer panes.[1]

Once all the checks had been completed, the spacecraft was reassembled and placed on temporary display outside of Hangar S for the official ceremony in which Scott Carpenter and Walt Williams were presented with NASA's Distinguished Service Medal by James Webb. Following this, the spacecraft was given a further extensive examination by McDonnell engineers and technicians before being handed over to the Manned Spacecraft Center in Houston, Texas.

© Springer International Publishing Switzerland 2016

C. Burgess, *Aurora 7*, Springer Praxis Books, DOI 10.1007/978-3-319-20439-0_8

As with John Glenn's *Friendship 7*, the compact capsule would be in high demand as a prime display object.

McDonnell engineers and technicians gave *Aurora 7* a detailed inspection before the spacecraft was released back to NASA. The craft's heat-faded logo can be seen on the right side. (Photo: McDonnell Aircraft Corporation)

A WELL-TRAVELED SPACECRAFT

Records seem to indicate that *Aurora 7* first went on public display in the "Man in Orbit" exhibit at the NASA-Plain Dealer Space Science Fair, Cleveland, Ohio, from November 23 – 2 December 1962. The *Cleveland Plain Dealer* newspaper for 24 November reported that it had taken four days for the spacecraft to complete the 1,100-mile road journey from NASA's Manned Spacecraft Center to the shores of Lake Erie. For exhibition purposes, the escape hatch of the spacecraft had been removed along with the astronaut's bulky couch so that the interior could be better viewed by the public. There was no protective covering over the spacecraft's exterior.

The scorched *Aurora 7* spacecraft was an extremely popular exhibit at the 1962 NASA-Plain Dealer Space Science Fair, located in the "Man in Orbit" area of the Fair. (Photos: Courtesy NASA Glenn Research Center)

It is next mentioned as a static display at the United States Space Park, a feature of the 1964–65 World's Fair. Held at Flushing Meadows Corona Park in Queens, New York, the Fair was designed to showcase mid-20th-century industry and technology, and a total of 51 million people passed through the gates in its two six-month seasons. The Space Park, set on two acres, was jointly sponsored by NASA, the Department of Defense, and the Fair Corporation. Among other exhibits were a full-scale model of the aft skirt and five F-1 engines of the first stage of a Saturn V, a Titan II booster topped with a mock-up Gemini spacecraft, and similarly an Atlas with a non-functional Mercury spacecraft and a Thor-Delta rocket.

The *Aurora 7* spacecraft was on open display at ground level, as were full-scale models of an X-15 aircraft, an Agena upper stage, a Gemini spacecraft, an Apollo command/service module, and what was then called a Lunar Excursion Module. Quite amazingly, *Aurora 7* was exhibited out in the open and not protected from the public in any way; patrons could (and did) run their hands across the historic shingled exterior.

On 26 October 1967, five years after it had carried Scott Carpenter on his orbital Mercury flight, NASA Langley Research Center in Hampton, Virginia, officially handed over *Aurora 7* to the Smithsonian Institution, where it was placed on display in the Air and Space Museum in Washington, D.C. The following year the historic spacecraft was loaned out for display in observance of Armed Forces Day at Robins Air Force Base in Georgia. It was on exhibit from 29 April to 13 May, when it was returned to the Smithsonian.

From 22 October 1968 to 1 June 1969, *Aurora 7* was loaned to the NASA Ames Research Center, Moffett Field, California, where it went on display. The spacecraft then traveled to the Pacific Science Center Foundation in Seattle, Washington, for the Center's "Man and the Moon" exhibit, on loan from 27 June 1969 to 1 September 1969.

The next beneficiary of the much-traveled spacecraft was the Naval Aviation Museum at NAS Pensacola, Florida, where it remained for three years from 15 December 1969 to 15 December 1972. It was then moved cross-country to Scott Carpenter's home town of Boulder, Colorado, where it went on display for the Department of Aerospace Engineering at the University of Colorado. Originally the spacecraft was scheduled to be exhibited at the university for three years from 18 March 1973, but this was later extended by another year to 18 March 1977.

Aurora 7 was then flown overseas to the Hong Kong Space Museum on the Kowloon Peninsula, under a sponsorship deal with NASA and the Smithsonian Institution, who had agreed to lend the spacecraft, a Moon rock, and spacesuit to the museum for exhibition. *Aurora 7* went on exhibit on 1 July 1982 and remained in the Hong Kong Space Museum for the next three years before being returned to Washington, D.C. It came back in what was described as a "poor condition," with the protective Plexiglas cover cracked in places and quite badly scratched. It was decided to carry out some minor refurbishment work on the spacecraft and re-cover it. This time Lexan thermoclear polycarbonate was chosen over Plexiglas for the operation.

On 1 April 1986, *Aurora 7* was delivered (now protected in its plastic shell) to the Henry Crown Space Center theatre, located within the Museum of Science and Industry, Chicago, Illinois. The original agreement called for it to be on display from July 1986 for three years.

At the end of this time the loan agreement was extended by a further three years, and it still remains there today on permanent display.[2]

Aurora 7 on permanent display in the Museum of Science and Industry, Chicago. (Photo: Richard Kruse, licensed under a Creative Commons Attribution-Noncommercial 3.0 United States License)

In a 2013 interview, Scott Carpenter was asked whether he had been to see the spacecraft. "I went back a couple of years ago and looked at it. I revisited my old friend, patted it on its shoulder and enjoyed the reunion. That machine and I went through a lot together. In that machine, I had a great experience. And I feel very fortunate to have had it."[3]

Today there is one lingering and baffling mystery concerning the *Aurora 7* spacecraft. It is evident from images taken post-flight that Cece Bibby's artwork on the side of *Aurora 7* made it through re-entry relatively unscathed, albeit heat faded. Somewhere in the following decades, however, her artwork did not remain with the spacecraft. When space historian Francis French of the San Diego Air and Space Museum met with curators at the Chicago Museum of Science and Industry in September of 2007 to closely examine the exterior of *Aurora 7*, only a very small amount of paint was evident on the shingles to either side of where the main artwork had been. The main shingle on which Cece Bibby completed her painting had been replaced with a plain shingle bearing no paint at all. The curators were mystified. French even asked Scott Carpenter about the painted shingle missing from his spacecraft, but he too was puzzled by this and had no explanation.

Where the *Aurora 7* artwork is now – whether lost, sitting on a dusty shelf in some building, or kept by an engineer as a souvenir – remains an unresolved mystery.

Scott Carpenter is reunited with *Aurora 7* at NASA's Manned Spacecraft Center, Houston, in 1962. (Photo: NASA)

LOSS OF A SPACE FLIGHT LEGEND

Malcolm Scott Carpenter died in hospice care in Denver, Colorado, on Thursday, 10 October 2013. He had succumbed to complications following a stroke suffered at his Vail home in September that put him in the Swedish Medical Center and, eventually, The Denver Hospice at Lowry.

"On Sunday, we watched the Broncos-Dallas game and he was very verbal," his daughter Candace ("Candy") Carpenter noted. But he took a turn for the worse the next day, and never recovered. He died peacefully at 5:30 a.m. with his wife Patty by his side.[4] "He wanted to be the best pilot, he wanted to be the best navigator, he wanted to be the best father, he wanted to be the best diver," Candy added. "He wanted to be the best at everything." She said he also came from an era when astronauts regularly put their lives on the line testing out new equipment and new technologies in situations and conditions humans had never before faced. "That was one of the things he was most proud of, was to be able to go to space and prove or disprove all those fears. He loved machines."

The loss of Scott Carpenter meant that only one member of the famed Mercury Seven selected back in 1959 was still around to speak of him, and there was sadness in his voice as he talked about the man from Boulder, Colorado, who also grew up with the nickname of Buddy. "Scott Carpenter was my lifelong friend," former astronaut and Ohio senator John Glenn said in a statement issued the day Carpenter died. "History books will remember him as an explorer of the heavens and the seas. Today I remember a statement Scott made over fifty years ago as I was launched into space. It was 'Godspeed, John Glenn.' These words meant a lot to me at the time and since, because I knew they were spoken from the heart, from our friendship and his concern for me and our mission. To paraphrase: 'Godspeed, Scott Carpenter, great friend.' You are missed."

As well as his wife, Patty Barrett Carpenter, he is survived by four sons: Robyn Jay, Matthew Scott, Nicholas Andre and Zachary Scott; two daughters, Kristen Stoever and Candace Carpenter; three step-children, one granddaughter, and five step-grandchildren. Two sons from his first marriage, Timothy Kit and Marc Scott, predeceased him.

On Saturday, 2 November, dozens of friends and family members and a number of Colorado politicians attended Carpenter's private funeral at St. John's Episcopal Church in Boulder, Colorado, followed by a public memorial. Colorado Governor John Hickenlooper had earlier ordered that flags be lowered to half-staff on all public buildings across the state that day as a mark of respect to honor the late astronaut. Several former astronauts also attended the service, some serving as honorary pallbearers. Present were Apollo astronauts Gene Cernan, Charlie Duke, Rusty Schweickart, Jack Schmitt and Dave Scott, as well as shuttle veterans Dan Brandenstein, Bob Crippen, Jim Reilly, Bruce McCandless, Dick Truly and Charlie Walker. Dee O'Hara, nurse to the original astronauts, and Suzan Cooper, the widow of original astronaut Gordon Cooper, also attended. The flag-draped casket was carried into the church as a bell tolled from the church tower.

Scott Carpenter, with his wife Patty by his side, reaches back to shake the hand of John Glenn at a 40th anniversary celebration of Apollo 11 on Capitol Hill, Washington, in 2009. Annie Glenn is obscured behind Carpenter. (Photo: Reuters/Larry Downing/Files)

At the memorial service, John Glenn said in his eulogy that his late friend possessed an adventuresome spirit and was driven to know everything he could about the universe. "Scott's curiosity knew no bounds," said Glenn. He recalled that Carpenter loved music, and had requested the hymn "Be Still My Soul" be sung at his funeral. With a sad smile, Glenn also reflected on a time when he and Carpenter had tried to harmonize on the song "Yellow Bird," with Carpenter later saying, "We weren't much good, but we were loud."

NASA Administrator Charles Bolden, himself a former astronaut, also attended the funeral. In a statement made soon after hearing of the astronaut's passing, he had paid tribute to Carpenter. "As one of the original Mercury 7 astronauts, he was in the first vanguard of our space program — the pioneers who set the tone for our nation's pioneering efforts beyond Earth and accomplished so much for our nation." At the funeral Bolden nominated Carpenter as a tireless explorer and an unforgettable character. Bolden pointed out that he was just a teenager when Carpenter orbited the Earth, and remembered "being tremendously moved by his bravery. Today," he added, "we bear witness as he soars once more into the heavens on his journey to eternity."

Following the service, four F-18 fighter jets flew the missing-man formation over the church against a cloudless autumn sky.

Patty Carpenter receives the flag that was draped over her husband's casket from U.S. Navy Rear Adm. Thomas Bond. (Photo: *collectSPACE.com*/Robert Pearlman)

Carpenter's daughter and co-biographer, Kris Stoever, later told *Outside* magazine, "In the spring, when the snow clears, we'll take his ashes to the Frye place, near Clark, Colorado. Homesteaded in 1901 by his great-uncle John, it hosts family camping trips when not under snow. The grave site will be consecrated among the aspens, near the headgate, and we'll inter his remains at 8,300 feet. This is what my father wanted, to be buried in a place that had shaped him."[5]

BY WAY OF TRIBUTE ...

Tracy Kornfeld is a talented American website designer and space flight enthusiast who set up the highly-popular and informative *scottcarpenter.com* website in cooperation with Scott Carpenter. As a result, they became friends and colleagues in the *Aurora 7* astronaut's latter years.

"I first met the Mercury Four (Scott Carpenter, Wally Schirra, Gordon Cooper and Bill Dana) no later than 2000. Bill Dana is a comedian whose hilarious astronaut skits in the 1960s saw him bestowed with the honorary title of 'The Eighth Mercury Astronaut.' During the course of that first introduction they all asked what I did for a living, which was designing websites for, at that time, the three-year-old internet (as we know it today). Scott was very interested in the technology and asked me to explain how it all worked. He thought that having a personal website would be a good idea at some point. The next time I saw him was in Washington, D.C. in 2002. I had a minor outline for a site drawn up for him, which he loved and wanted to discuss with his family. Sometime later, he called me to apologize that he wasn't going to use my design as his family wanted to go in a different route. That was fine with me. I was just happy that Scott showed an interest in my work.

"Later on, Wally Schirra – who had no interest in the internet – hired me to design his site and Bill Dana and I consulted about his History of Comedy website that he was working on for Emerson College in Boston. Scott stayed in touch with me and we became friends. I always enjoyed my conversations with Scott because we talked about everything *but* the space program. We had a common love of SCUBA diving and it turned out we had a lot of mutual diver friends. We also talked about his beloved Shelby muscle car and his love of anything technical. Our conversations fascinated me as Scott would wax lyrical about the topic of the day. I would tell people that Scott was a poet, not a pilot (or naval aviator in his case – I was always reminded that pilots guide ships into harbors).

"A few years later, his website was in disrepair and Scott contacted me to redo it. By the time I was done, he was averaging 50,000 visitors a month and we were constantly tweaking and updating it with video and interviews as well as a few opinion pieces. Scott was very pleased with it and I was always happy when the phone rang with Scott on the other end asking, 'How's my old friend, Tracy Kornfeld?'

"I was thrilled to see him happy with the design and launch of the limited edition 50th anniversary patch of his *Aurora 7* flight. He was very proud of it and was happy with the way his 50 years was presented on his site. I was honored to be invited to the 50th anniversary party in NYC and to rub elbows with Scott and other invited dignitaries.

"That was the final time I saw Scott, though we spoke on the phone all the time. Upon his passing, his website exceeded one million visitors over a three-day period. I think Scott would be proud that people were reading about his legacy the way he wanted it presented to the public via text, video and audio interviews. I felt that I had helped accomplish his mission.

"As of this writing, I still maintain Scott's website with his wife, Patty, and will keep it going as long as his family is interested. It is a living testament of a great man, a poet, that was written in his own words and told his true story."[6]

(Left): Tracy Kornfeld with Scott Carpenter and (right) the great man with Bruce Moody. (Photos courtesy of Tracy Kornfeld and Bruce Moody)

Another space flight aficionado with fond memories – and one in particular he wanted to share – is musician and songwriter Bruce Moody from South Carolina.

"I met Scott Carpenter in Seabrook, Texas in February 1990. I had become friends with one of the Shuttle Flight Trainers at Johnson Space Center in Houston. He and his training team used to come see my band when we were in town. It turns out space people want to be rock musicians and rock musicians want to be space people. This guy's dad, Bill Todd, was best friends with Scott when the two of them were growing up in Boulder, Colorado. In fact, there's a picture of him with Scott working on an old car in *Life* magazine when *Life* was doing a story about Scott's upcoming flight [issue 18 May 1962]. Well, they remained lifelong friends with each other, including Bill, Jr., who was my newfound friend. Bill used to invite me down to JSC on Wednesday nights, when I wasn't on the road, to play his 'victim' in the Space Shuttle simulator as he set up the next day's 'nits' and 'malfs' for the crew he was currently training. I've got some great video footage of us in the trainer while I'm trying to fix bad things going on and Billy explaining to me why I'm getting ready to make a new and very expensive hole in the ground!

"On one such Wednesday, Bill asked me come a little early, around 3 p.m. Now, anyone who's driven down I-45 South from downtown Houston at 3 p.m. can tell you it's slow going. I finally got to Billy's house in Seabrook, a modest little neighborhood just across from JSC where many astronauts lived with their families during their mission training. I went to the back door as usual, Billy answered and said, 'Hey, go have a seat in the den and I'll bring us some iced tea.' I walked into Billy's den and there's this short, gray haired guy sitting on the sofa. I looked at him and suddenly realized – hey! That's Scott Carpenter! Having met a number of 'famous' musicians in my life, I knew just how to act. I walked up to him, he stood up and he said, 'Hey, you're Bruce Moody, aren't you?!' He stole my line! I was dumbfounded!

"I finally managed to say, 'I've been wanting to meet you since I was about 6 years old!' and he said 'Thanks. You really know how to hurt a guy!' Immediately, he started asking me about my music, my band, the recording process and, of course, the wake-up song we were working on through Billy for STS-31. I couldn't get a word in edgewise about Mercury or *Aurora 7*. Finally, it was time to head over to the space center to learn more about fuel cells and to also learn why it's a bad idea to come up short at the HAC when you're lining up to land at KSC at the end of a Shuttle mission. The HAC is the Heading Alignment Cone, which is a big 300-degree turn that the shuttle makes to accomplish energy management.

"Over the years, Scott and I kept in touch sporadically, finally seeing each other in person at an autograph show in Washington, D.C. in 2002 or 2003, I believe. But my favorite time with Scott was the long weekend in 2008 when the two of us went to visit Cece Bibby. Cece was the artist who painted the insignias on the spacecraft of John Glenn (*Friendship 7*), Scott Carpenter (*Aurora 7*) and Wally Schirra (*Sigma 7*).

"Cece was living in a very nice retirement community in Hiawassee, Georgia while recovering from a recent stroke. Scott and I had discussed the possibility of going to visit Cece at some point if his schedule ever brought him to the Carolinas area. Scott called me one afternoon to say that he and Charlie Duke [Apollo 16] were doing an Omega watch joint appearance at a hotel in Winston-Salem, North Carolina, not too far from me and sort of in the area close to Cece. This seemed like a nice window of opportunity to accomplish our visit to Cece.

"After spending a week or so working out the details over the phone with Scott, it was decided that I'd pick him up at the Winston-Salem hotel on a Saturday morning at 7 a.m. and then the two of us would drive to Cece's place from there.

"Now, Hiawassee is in the mountains of Georgia, sort of tucked into the lower southwest corner of North Carolina and the northeast corner of Georgia. The route is filled with a lot of those 'you can't get there from here' roads. No worries. I had an 8-page MapQuest printout for our 5-hour drive and a list of questions memorized for my trapped guest of honor. As many of you know, it's nearly impossible to get Scott to talk about himself, so we talked about his early recollections of Cece, Dee, the Cape and, oh yes, the other astronauts, as they were exclusively known at the time. I will tell you all here and now that when I asked him if Gordo Cooper was 'the best pilot you ever saw,' Scott said, 'No, he wasn't the best. Probably of the seven of us, it'd be a tie between Al and Gus for pure pilot instinct and skill. Amongst the rest of us, it was probably a tie.' Wow! That only took 18 years and 3 hours of winding mountain roads, but it was worth it!

"The best part of the trip was just hanging out with Scott. He was so curious about learning. He really got into trying to identify the different trees and flowers along the beautiful North Carolina mountain roads. When we first arrived at Cece's place, she had told the staff that Mercury astronaut Scott Carpenter was coming. The center's staff met us in the lobby and took us to a private dining room for a special lunch. Cece was overjoyed to see Scott and you could just sense the fondness the two had for each other. Luckily, with Cece's stroke, her long-term memory wasn't affected at all at that time and the two of them reminisced for a long time. It was a great time to be a fly on the wall and enjoy my strawberry shortcake at the same time. Cece reminded Scott about the pranks Gordon Cooper played on her, and the big one she played on him (something about carrot cake), and the time Alan Shepard 'demanded' to drive her precious British race car and she turned him down. There was also the time Cece, having enough of being pushed around while she tried to get her work done on Scott's spacecraft painted a big red 'X' on Guenter Wendt's white overalls, or as Cece told it, imitating Guenter's German accent, 'Ven he vas trying to get me off ze gantry in ze high vinds dat ver blowingk in from ze coast!' Why didn't I record any of this?

Cece Bibby is reunited with Scott Carpenter in a 2008 visit. (Photo: Bruce Moody)

"But mostly, I will always remember Scott's kindness to Cece and to the other people that he met over that weekend. We took Cece out for dinner that night and for lunch the next day before we left. During our time at the center, there were a lot of older folks who remembered Scott's Mercury flight as well as that most interesting time in our nation's history. I was very young at the time of Scott's flight (7 years old), but I have a sense of what America was all about at that time and I realized that Carpenter was an Ambassador of that time to these people. He very patiently went through many of the questions, answers and stories he'd gone through thousands of times over the years for these people, just like it was the first time. He was an intriguing storyteller. He signed about 30 autographs. It just

Malcolm Scott Carpenter, renowned astronaut and aquanaut. (Photo courtesy of Hartriono Sastrowardoyo)

so happened that I brought along some photographs and Sharpies with me! When we left, Cece told me she now had 'bragging rights' for the next year! It was the happiest I'd ever seen her and it was great watching these two old friends make time stand still for a few priceless hours.

"I had to get Scott to the airport in Greensboro, North Carolina, the next day for a 4 p.m. flight to Chicago, so after lunch with Cece, it was back in the car for our now 4-hour journey out of the mountains and on to Greensboro. I did ask him about going to the Moon and to Mars and he told me that, at the time of his and John's flights, it was obvious we would get to the Moon and to Mars, but he thought it might not happen until the mid-1980's! When I asked who he would have liked to have gone to the Moon with, he said, 'Probably John or Gus [Grissom].' When I asked him why, he said, 'Because I would have wanted to come back to Earth when we were done.'

"Moments after dropping Scott off at the Greensboro airport, I phoned his daughter Kris in Colorado to let her know I'd safely delivered him there, that all was well, and that his sheen was still intact, which she thought was hilarious.

"Knowing Scott Carpenter was very easy. It was his kindness, his curiosity and his gentle nature that allowed him to interact with anyone and make them feel special. Those who came in contact with him instantly recognized that there was a rare gift of pure humanity in Malcolm Scott Carpenter.

"After Scott lifted off on his last trip from Earth to the heavens on October 10, 2013, I was reminded of what Daniel Patrick Moynihan had once said about JFK after that terrible day in Dallas. 'We will laugh again; it's just that we'll never be young again.'"[7]

It is left to Scott Carpenter himself to have the final say. When asked in 2012 what he thought his legacy might be, he had a typically uncomplicated notion of his role in history. "I carry it in my title, I guess," he replied. "I was an aquanaut and an astronaut. And that's good enough for me."[8]

REFERENCES

1. *Postflight Inspection Report* (10.1), from *Post-launch Memorandum Report for Mercury-Atlas No. 7 (MA-7)*, NASA, Manned Spacecraft Center, Houston, 15 June 1962
2. Spacecraft information courtesy Smithsonian Institution National Air and Space Museum Registrar's file, NASM 1851, A19680263000
3. Woody Paige interview with Scott Carpenter for *Mile High Sports* magazine, issue February 2013
4. John Aguilar, *Daily Camera* Boulder News article, "Boulder Astronaut Scott Carpenter Dies at 88," posted 10 October 2013
5. Kris Stoever, article "Scott Carpenter's Grand Adventure," *Outside* magazine online, 1 November 2013, available at *http://www.outsideonline.com/outdoor-adventure/exploration/Scott-Carpenters-Grand-Adventure.html*
6. Tracy Kornfeld, email correspondence with Colin Burgess, 30 August 2014 to 3 February 2015
7. Bruce Moody, email correspondence with Colin Burgess, 20–21 January 2015
8. Mitchell Byars, *Daily Camera* (Boulder, Colorado) newspaper article, *Boulder-bred astronaut Scott Carpenter – in town for park ceremony Thursday – reflects on spaceflight*, issue 19 September 2012

Appendix 1

MA-7 conference: Flight results in detail

As reported in NASA's Space News Roundup *bulletin: On Tuesday, 21 August 1962, some 1,500 scientists, engineers and technical personnel from the fields of industry and education as well as representatives of Congress, the President's Scientific Advisory Committee and various space science boards gathered in Houston's Rice Hotel ballroom for an all-day session in which seven reports on key aspects of the MA-7 mission were interspersed with discussion periods. Following are some of the reports given and discussed that day.*

THE PILOT: SCOTT CARPENTER

Carpenter added significant information for further space efforts and confirmed many of John Glenn's observations, including experiencing less noise and vibration than expected prior to and during powered flight; a lack of discomfort and improved mobility in a weightless state; the presence of the particles in space that Glenn called "fireflies" and Carpenter referred to as "snowflakes," observations of the bright "airglow" band on the horizon, and the success of a prolonged period of "drifting" flight.

"During the pre-launch period I had no problems. The couch was comfortable, and I had no pressure points. The length of the pre-launch period was not a problem. I believe I could have gone at least twice as long. Throughout this period, the launch vehicle was much more dormant than I had expected it to be. I did not hear the clatter that John Glenn had reported.

"I had expected to feel the launch vehicle shake, some machinery start, the vernier engines light off, or to hear the lox [liquid oxygen] valve make some noise, but I did not. Nothing happened until main engine ignition; then I began to feel the vibration. There was a little bit of shaking. Liftoff was unmistakable.

"About a minute and a half after liftoff, the sky changed in brightness rather suddenly. It was not black, but it was no longer a light blue. The noise and the vibration increased so little during maximum dynamic pressure that it would not be noticed unless you were looking for it. The booster engine cutoff was very gentle. Three seconds later, staging

© Springer International Publishing Switzerland 2016
C. Burgess, *Aurora 7*, Springer Praxis Books, DOI 10.1007/978-3-319-20439-0

occurred. There was no mistaking staging. Two very definite noise cues could be heard: one was the decrease in noise level that accompanied the drop in acceleration; the other was associated with staging. At staging there was a change in the light outside the window and I saw a wisp of smoke.

"At tower jettison, I felt a bigger jolt than at staging, and the tower was gone in a second. Out the window, the tower could be seen way off in the distance, heading straight for the horizon. It was rotating slowly with smoke still trailing out of the three nozzles.

"At SECO [sustainer engine cutoff] the rapid drop-off in acceleration was hardly noticeable. The best cues to the end of the powered flight were weightlessness and absolute silence."

The turnaround, Carpenter said, was just like the [Procedures Trainer] except that he was distracted by the onset of weightlessness. "Following the turnaround, I watched the expended launch vehicle through the window as it fell behind me, tumbling slowly. It was bright and easily visible. I could see what looked like little ice crystals spewing out the sustainer engine nozzle. They seemed to extend for two or three times the length of the launch vehicle, in a gradually broadening fan pattern."

The sensation of weightlessness was exactly as he had expected, Carpenter said. "It was very pleasant, a great freedom, and I adapted to it quickly. Movement in the pressure suit was easier and the couch was more comfortable. Later, when I tried to eat the solid food prepared for the flight, I found it crumbled in its plastic bag. Every time I opened the bag, some crumbs would come floating out; but once a bite sized piece of food was in my mouth, there was no problem. It was just like eating here on Earth."

In contrast to Russia's Gherman Titov, disorientation was not a problem. Carpenter stated that he knew at all times where the controls and other objects in the capsule were in relation to himself. At times, when the gyros were caged and nothing was visible out of the window, he had no idea where the Earth was in relation to the spacecraft, but he had only to wait and the Earth would once again appear in the window. The wider field of view of the periscope was particularly useful for Earth-spacecraft orientation, but Carpenter reported that without the periscope, the window would have been adequate.

Several unusual flight attitudes were investigated. "One of these was forward inverted flight. I think I could pick out the nadir point, that is, the ground directly below me, very easily without reference to the horizon. I could determine whether I was looking straight down or off at an angle. During portions of the second and third orbits, I allowed the spacecraft to drift. Drifting flight was effortless."

Carpenter noted that for normal maneuvering in orbit, fly-by-wire, low thrusters only, seemed to be the best system.

"At balloon deployment, I saw the confetti as it was jettisoned but it disappeared rapidly. The balloon … looked like it was a wrinkled sphere about eight to ten inches thick … [with] small protrusions at each side … Its motion following deployment was completely random.

"The view of Earth looked exactly like the pictures from other Mercury flights. The South Atlantic was 90 percent covered with clouds but all of western Africa was clear. Jungle areas showed up green."

Over California, Carpenter saw the area around El Centro quite clearly. "I saw a dirt road and had the impression that had there been a truck on it, I could have picked it out."

He did not see any more stars than could be seen from Earth, and said he was convinced that a lot more stars can be seen from the ground than through a spacecraft window. But the lights on the control panel, which could not be dimmed, had made it difficult to look out at the stars.

At dawn on the third orbit, Carpenter accidentally discovered the source of the previously reported "fireflies" when he inadvertently struck the hatch, dislodging a cloud of them which flew past the window.

After retrofire, in which he experienced several difficulties, Carpenter began to hear the hissing noise outside the spacecraft described by Glenn, and to see the re-entry glow of the tenuous upper atmosphere, heated by the spacecraft's passage. "I could see a few flaming pieces falling off the spacecraft." He noticed an orange glow about the window and a green glow around the cylindrical section of the spacecraft.

Carpenter deployed the drogue parachute manually at 25,000 feet after deciding that the spacecraft oscillations were getting too serious, and operated the main parachute deployment switch manually at 9,500 feet without waiting for automatic deployment. "The landing was less severe than I had expected … more noticeable by the noise than by the g-load. I was somewhat dismayed to see water splashed on the face of the tape recorder box immediately after impact. The spacecraft did not immediately right itself, listing halfway between pitch down and yaw left. I knew that I was way off my orbital ground track because I had heard earlier the Cape CapCom transmitting blind that there would be about an hour for recovery. I decided to get out at that time. Egress is a tough job. The space is tight and the small pressure bulkhead stuck slightly."

Carpenter then described his activities during the time until he was spotted and recovery took place. He was dunked in the sea during the hoisting operation which took him aboard the rescue helicopter, and once aboard took off his boots and "poked a hole in the toe of my left sock and stuck my leg out the window to let the water drain out of my suit."

FLIGHT OPERATIONS CREW REPORT

"The results of the MA-7 flight provide additional evidence that man is ready for a more extended mission in a weightless environment," personnel of the Flight Crew Operations Division stated in their report to the MA-7 Results Conference.

This report on the pilot's performance was prepared by Helmut A. Kuehnel, William O. Armstrong, John J, Van Bockel and Harold I. Johnson. It was one of seven reports presented at the conference.

They reported that Scott Carpenter spent more than 70 hours in the ALFA [air-lubricated free-attitude] trainer and Mercury spacecraft Procedures Trainer prior to the flight, doing 73 simulated missions, 143 simulated failures, and 255 simulated control maneuvers. He spent 45 hours in spacecraft systems checks, almost twice as much as John Glenn. In addition, Carpenter spent 80 hours in the MA-6 spacecraft while acting as backup pilot for Glenn during the checkout period at the launch site.

Investigating a pilot's ability to observe an object in space, "Carpenter readily sighted the detached sustainer stage after turnaround and calculated the distance at about 300 yards. He continued to observe and photograph the sustainer for 8.5 minutes, at which time it was about three miles away."

A continuous period of 1 hour and 6 minutes in drifting flight during the third orbital pass was reported as not at all disturbing. Data showed the spacecraft attitude rates were less than 0.5 degree per second during one period of the drifting flight, so little that the Moon stayed at or near the center of the window for a significant period.

A malfunction of the pitch horizon scanner circuit required the pilot to manually control the spacecraft's attitude during retrofire, and the maneuver was believed to have proceeded normally except for late ignition of the retrorockets. However, the spacecraft overshot the intended landing point by about 250 miles. The pilot had backed up the automatic retrofire system by pushing the manual button when the event did not occur. Retrofire occurred three to four seconds late, accounting for about 15 to 20 miles of the total overshoot error.

Radar tracking data indicated that the mean spacecraft pitch attitude during most of the retrofire period was essentially correct. Some deviations showed up in roll attitude, but roll errors of this size would have a negligible effect on landing point dispersion. Thus the error in landing position resulted primarily from a misalignment in spacecraft yaw attitude which, as deduced from radar tracking data, was believed to have been about 27 degrees during the retrofire maneuver.

GROUND COMMUNICATION

"The Mercury Network performed very well in support of the Mercury-Atlas-7 mission," Goddard Space Flight Center scientists told participants at the conference. The report was authored by James J. Donegan and James C. Jackson of the Manned Space Flight Support Division.

"No problems were encountered with computing and data flow. The computers at Goddard accurately predicted the 250-mile overshoot immediately after … tracking data … were received. Radar tracking was generally horizon-to-horizon and the data resulted in accurate orbit determination early in the mission.

"Ground communications network performance was generally better than for the MA-6 mission. But the ground-to-spacecraft communications were slightly inferior, particularly when patched onto the conference network to allow monitoring by other stations." As with the MA-6 flight, telemetry reception was good.

For the MA-7 flight there was no mid-Atlantic ship, and the Indian Ocean ship had been repositioned in the Mozambique Channel.

The telemetry reception of aeromedical data on astronaut heartbeat rate, respiration, ECG, blood pressure and body temperature was good, and from horizon-to-horizon at all tracking stations.

The expected blackout of telemetry occurred at re-entry, when ionized air enveloped the spacecraft and blocked radio transmission. Although this effect began at a ground-elapsed time of 4 hours, 43 minutes and 58 seconds, telemetry contact was regained 4 minutes and 49 seconds later and held for about 12 seconds at the Grand Turk Island station. After that, the final loss of telemetry was due to extreme range and low elevation angle.

The dual ground command system operated normally in spite of several minor malfunctions that had no effect on the mission.

The teletype and voice network between ground tracking stations performed well with only three minor difficulties occurring, all of them non-critical interruptions.

HF and UHF voice transmission between ground and spacecraft was adequate, although the majority of stations reported a lower signal level than was experienced for MA-6.

MISSION OPERATIONS REPORT

Assistant Chief for Flight Control John D. Hodge and two other members of the Flight Operations Division, Eugene F. Kranz and William C. Hayes, were co-authors of a report on mission operations presented at the conference.

Launch vehicle countdown and the pre-launch phase were nearly identical to MA-6. The launch phase proceeded almost perfectly, the report said, and the powered flight was normal. "The flight was satisfactorily monitored by Mercury Network ground stations and no major flight discrepancies were evident until just prior to retrofire, when it was discovered that the automatic control system was not operating properly."

Some small changes from the MA-6 operational support were made, mostly associated with the development of support procedures for further missions of longer duration.

The flight plan was basically the same as for MA-6 with two differences: the astronaut had more manually controlled tasks to perform and a larger number of experiments.

A new network countdown was used, and a high degree of confidence in the countdown format established. The second part of the split countdown, that part held just prior to launch, was "probably as close to perfect for the launch vehicle, spacecraft and network as could ever be expected," in the words of the report.

A remote facility for transmitting air-ground voice to the Mercury Control Center through the Bermuda site transmitters was used for this mission, allowing the Mercury CapCom to transmit data to the astronaut in real time, thereby eliminating much of the requirement for relaying information between the Canaveral and Bermuda controllers.

MERCURY PROJECT OFFICE

"The performance of the Mercury spacecraft and the Atlas launch vehicle for the orbital flight of Astronaut M. Scott Carpenter was excellent in nearly every respect," reported John H. Boynton and E. M. Fields of the Mercury Project Office.

"All primary mission objectives were achieved. The single mission-critical malfunction involved a random failure in the pitch horizon-scanner circuit.

"This anomaly was compensated for by the pilot in subsequent operations so that the success of the mission was not compromised. A modification of the spacecraft control-system thrust units was effective. Cabin and pressure suit temperatures were moderately excessive but not intolerable.

"Some uncertainties in the data telemetered from the bioinstrumentation prevailed at times … however, associated information was available which indicated the continual well-being of the astronaut."

The report noted that the performance of the heat protection system was quite satisfactory. The entire group consisting of separation devices, rocket motors, landing system and internal spacecraft structure functioned satisfactorily. The ignition of the retrorocket motors was about three seconds later than expected but the performance of the motors was satisfactory. As planned, Carpenter manually deployed the drogue parachute when he felt the need for additional spacecraft stabilization. He also manually deployed the main chute at an altitude of about 9,500 feet rather than waiting for automatic deployment, which would have occurred at about 8,200 feet.

The spacecraft's severe list angle after landing was attributed to Carpenter's egress activity, which took place before the spacecraft would normally have righted itself to the erect position. The majority of sea water found in the spacecraft was believed to have entered through the small pressure bulkhead when Carpenter left the spacecraft. Small leaks in the internal pressure vessel would have contributed little of the water found in the cabin. The drops Carpenter noticed on the tape recorder immediately after landing were believed to have resulted from a surge of water which momentarily opened a spring-loaded pressure relief valve in the top of the cabin.

Of the control system, the report noted that post-flight analysis of the faulty pitch horizon scanner circuits was impossible since the scanners were lost when the antenna canister was jettisoned during normal landing procedure. The failure was believed to have been in the scanner circuitry and of an apparently random nature.

"Because the malfunctioning scanner circuit resulted in pitch errors in the attitude-gyro system, the pilot was required to assume manual control during retrofire," the report said.

"Double authority control was inadvertently employed at times during the flight and the high thrust units were accidentally actuated during certain maneuvers, both of which contributed to the high usage rate of fuel. MA-8 and subsequent spacecraft will contain a switch which will allow the pilot to control the use of the high thrust units."

High cabin temperatures during the MA-7 flight were attributed to a number of factors, such as the difficulty of selecting the proper water flow rates and a vulnerability of the heat exchanger to freezing-blockage when too high rates of water flow were used. Tests should determine whether the cabin temperature could be lowered without requiring a substantial redesign of the cooling system.

Concerning the high suit temperatures, "It is believed that the suit-inlet temperature could have been maintained in the 60 to 65 degree range … had not the comfort control valve been turned down early in the flight." Carpenter reduced the setting of this valve when the cabin heat exchanger indicated possible freezing.

The failure of the retrorockets to fire automatically from the clock was attributed to the fact that the pitch attitude gyro – which was shown to be in error – did not indicate that the pitch attitude was within acceptable limits. The attitude-permission circuits therefore would not pass the retrofire signal from the clock. Carpenter had waited for two seconds before he actuated the manual retrofire switch. An additional one-second delay before the ignition of the motors remained to be explained.

The cause of the failure of the balloon to inflate was attributed to a ruptured seam in the balloon skin. Aerodynamic drag measurements were thus invalid, since the surface area of the balloon was not known. In addition, the visibility portion of this experiment suffered because only two of the five surface colors on the balloon were visible to the pilot.

A heavy cloud cover over Woomera had effectively foiled the ground-flare visibility experiment on the first orbital pass, and it was discontinued for the remainder of the flight because the cloud cover remained. The exercise, the report stated, would be repeated on a future flight.

GODDARD SPACE FLIGHT CENTER

Two scientists from the Goddard Space Flight Center discussed the airglow layer at the Earth's horizon, the space particles reported by astronauts John Glenn and Scott Carpenter, and the flattening of the solar image at sunset in a paper presented by Dr. John A. O'Keefe and Winifred Sawtell Cameron, both of Goddard's Theoretical Division.

This 'Space Science Report' said that the airglow layer, brightest about 10 or 15 degrees above the horizon, was estimated as extending from 90 to 118 kilometers above the surface. Carpenter noted that it was relatively bright compared to the moonlit horizon and to Phecda, the star of 2.5 magnitude more formally catalogued as Gamma Ursae Majoris. Wavelength observations of the layer were made with a special filter supplied by Goddard, and the band was definitely identified as the 5,577 Angstrom layer.

Carpenter estimated the height of the layer to be from 8 to 10 degrees (Glenn's estimate was 7 degrees), and about twice the height of the twilight layer, which he estimated at five times the apparent diameter of the Sun. He also observed Phecda as it passed the middle of the band. But calculations showed the lower boundary of the layer to be at 73 kilometers, or about 2 or 3 degrees above the horizon, thereby illustrating the well-known illusory effect that exaggerates angles near the Earth's horizon.

The authors of the report noted that astronaut Gus Grissom may have observed the airglow layer during the daytime. He reported a grayish band at the top of the blue sky layer during his flight.

Carpenter did not note any vertical or horizontal structures in the layer, nor did he attempt a continuous survey around the horizon; however he did note the layer at several points along the horizon and believed it to be continuous.

Carpenter also noticed and photographed white objects like "snowflakes" at sunrise and at various periods following sunrise on all three orbits. Shortly prior to retrofire, the astronaut accidentally struck the spacecraft hatch when reaching for an instrument and promptly saw a cloud of particles fly past the window. He further struck the spacecraft walls, producing the same results each time. Two plausible sources within the spacecraft were "snow" formed by condensation of steam from the life support systems; and small particles of dust, waste, bits of insulation and other sweepings. The strongest theory was that the snowflakes formed by the freezing of water vapor from the spacecraft. The condensation probably occurred in the space between the heat shield and the large pressure bulkhead of the spacecraft (rather than outside) from the steam exhaust from the life support system. The snowflakes might have escaped into space through the ports, driven outward by expanding water vapor.

New information regarding refraction by the Earth's atmosphere of celestial objects seen from space was provided when astronauts Glenn and Carpenter obtained photos of the setting Sun that illustrate the effect strikingly. Carpenter recognized the phenomenon visually, but Glenn did not.

AEROMEDICAL OBSERVATIONS

"A review of the detailed medical examinations accomplished on two astronauts who each experienced approximately 4.5 hours of weightless space flight reveals neither physical nor biochemical evidence of any detrimental effect. Such flights appear to be no more physiologically demanding than other non-space-related test flights."

Thus opened the report entitled 'Clinical Aeromedical Observations' by Dr. Howard A. Minners, astronaut flight surgeon for MA-7; Dr. Stanley C. White, chief of the Life Systems Division; Dr. William K. Douglas, Air Force Missile Test Center, Patrick AFB, Florida; Dr. Edward C. Knoblock, Walter Reed Army Institute of Research, Washington, D.C.; and Dr. Ashton Graybiel, U.S. Naval School of Aviation Medicine, Pensacola, Florida. The report was presented by Dr. Charles A. Berry, chief, Aerospace Medical Operations Office.

"The experience gained in the MA-6 flight altered the medical planning for the MA-7 flight in two important aspects. A comprehensive medical evaluation ... was conducted at the earliest opportunity after landing when the pilot's impressions were freshest ... The flexibility of the procedure at the debriefing site was increased to take greater advantage of any medical symptoms which might appear."

The triple purpose of the clinical observations, the report said, was to determine astronaut fitness for the flight, to provide baseline information for the aeromedical flight controllers, and to measure any changes which might have occurred during the flight. A special diet was used for 19 days before the flight, and a three-day low residue diet immediately prior to the flight. Carpenter maintained his physical condition with daily workouts on the trampoline and with distance running.

Breakfast on the morning of the flight consisted of filet mignon, poached eggs, strained orange juice, toast, and coffee. The night before the mission, Carpenter gained about three hours of sound sleep.

Physical examination aboard the aircraft carrier after the flight showed the pilot without injury and in good health. He showed a mild reaction to the adhesive tape used to keep the four ECG sensors in place and at the blood-pressure microphone location.

Two special tests were used during both MA-6 and MA-7 to measure any effect of space flight and weightlessness on the human vestibular apparatus. In the first, the subject's ear was irrigated for 45 seconds with water below body temperature which could be warmed or cooled under precise control, and the temperature at onset of nystagmus [fine eye jerk] was noted. Neither astronaut showed a significant change in threshold temperatures before and after the flight. The other test measured the subject's ability to balance on successively more narrow rails, similar to those of a railroad track. Both astronauts showed a small increase in their post-flight versus pre-flight scores in this test, thus showing no detrimental change in balancing ability.

In addition, a xylose tolerance test was carried out to measure intestinal absorption during weightless flight, but the results were not as conclusive as in the Glenn flight.

The second portion of the medical report concerned physiological responses of the astronaut, and was written by Dr. Ernest P. McCutcheon, Dr. Charles A. Berry, and Robie

Hackworth, all of the Aerospace Medical Operations Office, MSC; Dr. G. Fred Kelly, USN; and Rita M. Rapp of the Life Systems Division.

"The heart-rate response to nominal exercise demonstrated a reactive cardiovascular system. An abnormal electrocardiogram (ECG) tracing was recorded during re-entry and is believed to have resulted from the increased respiratory effort associated with straining to continue speech during maximum re-entry acceleration. No disturbing body sensations were reported as a result of near-weightless flight. Astronaut Carpenter felt that all body functions were normal.

"Solid foods can be successfully consumed in flight, but precautions must be taken to prevent crumbling. The biosensors provided useful ECG data, with minimal artifact. The respiration rate sensor provided good pre-launch but minimal in-flight coverage. Because of erratic amplifier behavior, the rectal temperature thermister gave invalid values for approximately one-third of the flight. At the present time, the in-flight blood pressure cannot be interpreted.

"The instability of the body temperature readout is believed to have been the result of erratic behavior of the amplifier from 59 minutes to two-and-a-half hours after launch," or about one-third of the flight. Values at other times were considered valid.

"Carpenter stated that the flight was not physically stressful. He was subjectively hot and perspiring during the second orbital pass and the first half of the third pass, but was never extremely uncomfortable.

"Violent head maneuvers within the … helmet were performed several times without symptoms of disorientation or vertigo. Vision was normal throughout the flight, and colors and brightness of objects were clear and easily discernible. Distances were estimated by the relative size of objects. There was no detectable change in hearing. Somatic sensations were normal and no gastrointestinal symptoms were apparent.

"During flight, astronaut Carpenter consumed solid food, water, and a xylose tablet without difficulty. The solid food was in the form of bite size, three-quarter-inch cubes with a special coating packed loosely in a plastic bag and stored in the equipment kit. Because some crumbling was reported when he first attempted to eat, it is believed that the food was inadvertently crushed during final spacecraft preparation on the launch pad. The special coating having been broken, the food continued to crumble during flight. The pilot stated that the floating particles within the spacecraft were a potential inhalation hazard. Finally, the elevated cabin temperature caused the candy to melt. He reported the only difficulty was in getting the crumbled food particles to his mouth. Once in the mouth, chewing and swallowing of both solids and liquids were normal. Taste and smell were also normal.

"A total of 1,213 cc of water was consumed from the mission water supply. An estimated 60 percent was consumed in flight and the remainder after landing.

"Calibrated exercise was performed without difficulty at 03:59:29 [ground elapsed time]. Because of the overheated condition of the pilot, earlier scheduled exercises were omitted. A hand-held bungee cord with a 16-pound pull through a distance of 6 inches was used. Use of this device for a short period caused an increase of 12 beats per minute in heart rate with return to previous values within 1 minute. The heart-rate response to this nominal exercise demonstrated a reactive cardiovascular system.

"Attempts to produce autokinesis [a visual perception phenomenon where a stationary, small point of light in an otherwise dark or featureless environment appears to move due to involuntary eye muscle movements] were made on two occasions. Autokinesis was not produced but the tests were inconclusive."

REFERENCE

1. All reports were taken from NASA *Space News Roundup* bulletin, Vol. 1, No. 22, 22 August 1962.

Appendix 2

Scientific instruments carried on MA-7

Reproduced from: Post-launch Memorandum Report for Mercury-Atlas No. 7 (MA-7), Part 1: Mission Analysis. NASA Manned Spacecraft Center, Houston, TX, 15 June 1962.

35 MM HAND HELD CAMERA

A 35 mm Robot Recorder 36 was provided. After being lightened, a pistol grip handle was added along with other modifications to permit ease of operation by a spacesuited astronaut, and a clip was provided for attachment to the chart holder in weightlessness. It was equipped with a standard back assembly and a 30-foot film capacity magazine. Additional equipment included two interchangeable lenses, one a 75 mm, f/3.5 lens and the other a 45 mm, f/2.3 lens. Each lens system was provided with a UV-17 filter. This camera functioned well throughout the flight. Although the large capacity back reduced film changing to a minimum, it was still necessary to change films to accomplish specialized photography.

FILM

The 30-foot magazine was preloaded with Eastman Color Negative film (stock number 5250) and attached to the camera prior to insertion into the spacecraft. This film load represented a 250-exposure capability. The Massachusetts Institute of Technology provided a preloaded film (Eastman stock number SO-1030) for the horizon definition photography task. This film load provided approximately 70 exposures. The Weather Bureau experiment required a 36-exposure film load that was alternately spliced from Tri-X and infra-red film stocks. Also included was one 36-exposure roll of Ansco Super Hypan film to provide an alternate to the ECN for photographing the "fireflies" reported by John Glenn.

© Springer International Publishing Switzerland 2016
C. Burgess, *Aurora 7*, Springer Praxis Books, DOI 10.1007/978-3-319-20439-0

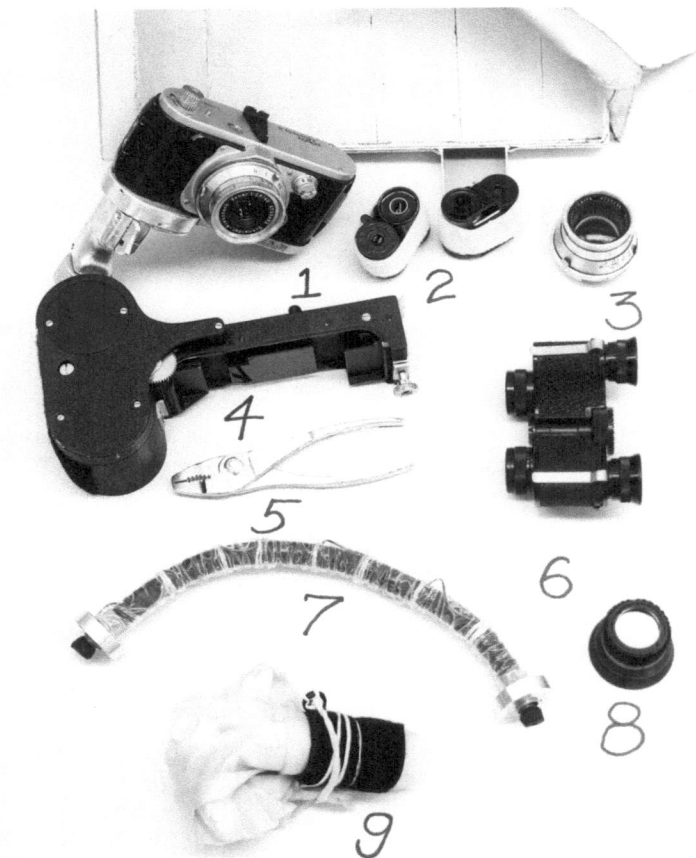

A specially made case housed the following equipment used by Scott Carpenter on the MA-7 flight. 1: Robot camera. 2: Extra film. 3: 75 mm telephoto lens. 4: 30-foot film for camera. 5: Pliers. 6: Binoculars. 7: Exercise device. 8: Filter for checking airglow. 9: Watertight bag for housing equipment in case of water egress. (Photo: NASA)

FILTER MOSAICS

Two filter mosaics were provided. These mosaics were mounted in holders designed to be inserted into the camera at the film plane. One was to be used with the MIT film and the other with the Weather Bureau film. The MIT mosaic consisted of two equal sections of Wratten filter numbers 29 and 47B. The Weather Bureau mosaic had five equal sections containing Wratten filter numbers 0.8 neutral density, 25, 47, 58, and 87. Of these, only the MIT mosaic was used and it performed satisfactorily.

Carpenter studies the modified camera that he would use on MA-7. (Photo: NASA)

PHOTOMETER

This device was the same one used during the MA-6 mission. It was used to view sunrise and sunset, to evaluate the pilot's capability to orient to the horizontal, and as a high and low level light meter. It was used by the pilot with satisfactory results.

BINOCULARS

The pilot was provided with a miniature pair of 8 × 20 binoculars. Clips were provided to permit attachment to the chart holder during orbit. The pilot reported utilization during flight was difficult due to the viewing angle with the window.

EXTINCTION PHOTOMETER

This device consisted of a calibrated, circular, varying density filter in a suitable mount. It was used on several occasions during the flight with satisfactory results.

AIRGLOW FILTER

This was the same device as used on the MA-6 flight, with a modified mount. It passed light at the 5577A wavelength. This device was used to view the airglow layer on the nightside of the Earth.

NIGHT ADAPTATION EYE MOVER

This device fitted the eye socket in such a manner as to prevent any direct light from reaching the eye. It was provided with a red lens to allow the pilot the use of his left eye during the adaption period. It functioned properly during the flight, but complete dark adaptation was prohibited by stray light within the spacecraft.

MAP BOOKLET, STAR NAVIGATION DEVICE AND INSERTS, AND FLIGHT PLAN CARDS

The pilot reported only that the glare from the star navigation device made it difficult to use. The balance of this equipment and its stowage was adequate.

EQUIPMENT STOWAGE

All equipment had female Velcro applied to strategic points, whereas male Velcro was applied to the stowage areas. Four equipment areas were provided in the MA-7 spacecraft. During the launch, retrofire and re-entry phases, equipment was stowed in three locations. First, the equipment container located to the pilot's right, below the hatch, held the 35 mm camera and its associated accessories; photometer, binoculars, and extinction photometer. Second, the "glove compartment" located in the left central section of the center instrument panel console contained the exercise device, film, filter mosaics, airglow filter, and the night adaptation eye cover. Third, the chart holder located below the periscope contained the map booklet, star navigation device and inserts, and the flight plan cards. During the orbital phase, the equipment was stowed either in these locations or on the Velcro applied to the hatch for this purpose. The pilot reported no difficulties with the stowage of any of the equipment.

Appendix 3

Results of MA-7 tethered balloon experiment

This experiment was designed to provide orbital observations of nearby objects of varying surface finishes and to measure the drag of an object of known aerodynamic characteristics in a region of free molecular flow. Balloon drag could then be related to atmospheric density and thus provide a density profile over the altitude range encompassed during the Mercury orbital pass.

The experiment was also intended to obtain qualitative information on the capability of the astronaut to estimate separation distance between the spacecraft and an object of known size and shape in space. The visual portion of the experiment was to evaluate the relative merits of various colors and surface finishes for optimum visibility at varying distances in a space environment. Additional objectives of this test include observations of the general stability qualities and damping characteristics of the tethered balloon. The appearance, brightness, and behavior of small diffuse reflecting discs were to be observed to provide a comparison with other foreign particles in space where appropriate.

The balloon was deployed in orbit at 01:38:00 by firing a squib. A small compressed spring then ejected the balloon and an inflation bottle from the container, along with two balsa block liners and the Mylar discs. The balsa blocks were semi-cylindrical in shape and about 6 inches long and 3 inches wide. One was coated with Day-Glo orange and black and the other with Day-Glo yellow and black. The Mylar discs were coated with aluminum foil on one side and a diffuse reflecting material on the other.

The balloon was tethered to the spacecraft by a 6-pound-test nylon line measuring 100 feet in length which was deployed from a spinning reel. The line was stripped from the reel when the balloon was fully deployed, but remained attached to a small strain gauge mounted in the bottom of the balloon container. Continuous strain gauge measurements were to be recorded on board the spacecraft until the drag test was completed, whereupon the balloon would be jettisoned and the rate and distance of separation between the spacecraft and balloon would be estimated by the astronaut.

© Springer International Publishing Switzerland 2016

C. Burgess, *Aurora 7*, Springer Praxis Books, DOI 10.1007/978-3-319-20439-0

Unfortunately, the balloon did not inflate completely and did not jettison. Therefore, drag measurements and rate and distance of separation of the balloon from the spacecraft were not obtained.

TEST RESULTS

At balloon deployment, the astronaut reported seeing the Mylar discs spread out and quickly disappear. His first impression was that the balloon had broken free from the spacecraft, but the object he was tracking was one of the balsa blocks. He observed this block for about 20 seconds, then the partially inflated balloon came into view. These observations were verified by pictures taken by the astronaut. Although attitude rates were noted in all three axes during and after deployment during the towed phase, the effect of these rates were unable to be conclusively determined.

Pilot comments and photographs showed that the balloon shape tended to be irregular and oblong, and appeared to be about 6 to 8 inches in cross section. The astronaut described the balloon motion as being completely random in nature. These random motions may have been caused by large changes in the attitude of the spacecraft after deployment. However, uneven aerodynamic loads likely existed on the irregular balloon shape and could also be expected to contribute to this random motion.

Onboard comments by the astronaut did indicate, however, that during the portion of both the second and third orbital passes when dynamic pressure was increasing, balloon motions tended to become more stable. Approximately 35 minutes after deployment, the astronaut initiated a series of control maneuvers to check the spacecraft control system, and the strain-gauge measurements indicated that fouling of the tethering line occurred during this period. This conclusion is further substantiated by the fact that subsequent spacecraft maneuvers were not registered on the strain-gauge system.

At approximately 03:14:00, the astronaut attempted balloon jettison, but the balloon did not release from the spacecraft. However, the onboard strain-gauge recording indicated a drop in gauge output from the level it had held since probable fouling to the output level of the unloaded gauge. This drop, which constituted a change of only 2 to 3 percent in gauge output, does provide a positive indication that the jettison squib fired and that the tethering line was severed.

Only the Day-Glo orange and the uncoated aluminum foil were visible to the astronaut, and these were the only colors which appeared in the photographs. Therefore, an effective evaluation of the colors could not be made on this flight.

SUMMARY

An analysis of the experimental results indicated that the balloon deployment, jettison, and instrumentation systems functioned satisfactorily during flight. Since the balloon failed to inflate properly at deployment, no useful drag and visual observation data were obtained. High rates of change in spacecraft attitude after balloon deployment, as well as the irregular shape of the partially inflated balloon, probably accounted for the random motion of the balloon observed during flight. Effective evaluation of the various colors was not possible since only part of the balloon was exposed.[1]

RESULTS OF FLUIDS IN SPACE EXPERIMENT: 50 YEARS ON

A half a century ago, only four months after John Glenn's historic *Friendship 7* flight, the first U.S. fluid physics space investigation took place aboard *Aurora 7*, the second orbital Mercury flight made by Scott Carpenter. This experiment addressed a critical question for human space flight. How do fluids behave without the presence of Earth's gravity? Would fluids break up into small droplets and disperse throughout the container in reduced gravity? Or would capillary forces prevail and spread the fluid across all of the container surfaces? Capillarity is the tendency of a liquid in contact with a surface to flow as a result of surface tension. How would spacecraft maneuvers such as attitude control or docking influence the behavior and location of fluids? Even small disturbances in a low gravity environment might cause the fluids to move or shift their position.

The answers to these questions were needed to determine the location of propellants in rocket vehicle tanks, especially cryogenics like liquid hydrogen and liquid oxygen, as well as other liquids needed for life support, such as water.

"These were new questions to spacecraft engineers in the late 1950s," said Bob Green of the NASA's Glenn Research Center. "Prior to this, engineers depended on gravity to 'mask' these capillary effects and reliably position any liquid at the bottom of a container, such as fuel in an aircraft fuel tank or even in your car's gas tank."

Researchers from the NASA's Lewis Research Center, now known as the Glenn Research Center, devised an experiment for Scott Carpenter to test the effects of capillarity and a piece of hardware to control the motion of the liquid in a tank, called a passive baffle. This used a simple cylinder, or standpipe, as a baffle inside a spherical container, which was partially filled with a liquid containing a green dye and a wetting agent. A video camera recorded the motion of the liquid. "The liquid behaved just as the Lewis scientists predicted," said Bill Masica, an original investigator and former Chief of the Space Experiments Division there. "In 1962, it certainly went a long way toward helping eliminate the mystery of how fluids would behave under weightless conditions."

The Mercury investigation was conceived with primarily liquid propellant management in mind, according to Mark Weislogel, Professor at Portland State University, and a principal investigator for fluid physics investigations currently aboard the International Space Station. "Its successful demonstration added confidence to design engineers in the emerging field of large-length scale capillary flow aboard spacecraft."

Early investigations like this one led scientists over the ensuing half century to perform further gravity-dependent fluid physics studies in space aboard sounding rockets, Apollo, Skylab, Space Shuttle, the Russian Space Station Mir, and the International Space Station. These experiments contributed to better understanding of the behavior of liquids in space under varying gravity conditions. Resulting data, detailing the behavior of liquids, led to design improvements for containers and transfer equipment to counteract the challenges triggered by microgravity, acceleration, and changes in direction.[2]

REFERENCES

1. NASA paper, "Second United States Manned Three-Pass Orbital Mission (Mercury-Atlas 7, Spacecraft 18)," section "Description and Performance Analysis," Edited by John H. Boynton, NASA Manned Spacecraft Center, Houston, Texas, February 1967
2. Linda Nero, International Space Station and Human Health Office, "Capillarity in Space – Then and Now, 1962–2012," NASA Glenn Research Center, 24 May 2012.

Appendix 4

Sequence of events during the MA-7 flight

Event	Pre-flight predicted time hr:min:sec	Actual time hr:min:sec
Booster engine cutoff (BECO)	00:02:10.1	00:02:08.6
Tower release	00:02:32.2	00:02:32.2
Escape rocket firing	00:02:32.2	00:02:32.2
Sustainer-engine cutoff (SECO)	–	00:05:09.9
Tail-off complete	00:05:05.3	00:05:10.2
Spacecraft separation	00:05:06.3	00:05:12.2
Retrofire sequence initiation	04:32:25.6	04:32:36.5
Retro (left) No. 1	04:32:55.6	04:33:10.3
Retro (bottom) No. 2	04:33:00.6	04:33:15.3
Retro (right) No. 3	04:33:05.3	04:33:20.5
Retro assembly jettison	04:33:55.6	04:34:10.8
0.05 g relay	04:43:55.6	04:44:44
Drogue parachute deployment	04:50:00.6	04:50:54
Main parachute deployment	04:50:37.6	04:51:48.2
Water impact	04:55:22.6	04:55:57
Main parachute jettison	04:55:22.6	04:56:04.8

© Springer International Publishing Switzerland 2016
C. Burgess, *Aurora 7*, Springer Praxis Books, DOI 10.1007/978-3-319-20439-0

Appendix 5

MA-7 capsule fuel consumption chart
(Times quoted are mission elapsed time)

Time	Mission Phase	Automatic System Fuel Used (lbs.)	Fuel Left (lbs.)	Manual System Fuel Used (lbs.)	Fuel Left (lbs.)
00:00:00	Launch	0	35.0	0	24.9
00:08:00	Turnaround and damping	1.6	33.4	0	24.9
01:33:32	First pass	15.8	17.6	8.5	16.4
03:07:04	Second pass to retro	5.8	11.8	5.1	11.3
04:33:21	Retro to 0.05 g	5.4	6.4	10.3	1.0
04:44:44	0.05 g to drogue	5.0	1.4	1.0*	0
04:50:54	Drogue to main	1.4*	0	0	0

*Fuel depletion occurred during this period

C. Burgess, *Aurora 7*, Springer Praxis Books, DOI 10.1007/978-3-319-20439-0

Appendix 6

Chronological summary of post-landing events

E.S.T. hr:mn	Elapsed time from landing hr:mn	Event
11:18 a.m.	–	As a precautionary measure, Air Rescue Service SC-54 was launched from Roosevelt Roads, Puerto Rico, to take station on the downrange end of Area H. It carried a specially trained pararescue team.
12:22 p.m.	–	Retrorockets were ignited.
12:33 p.m.	–	Calculated landing position was reported as being 19°24′ N. latitude, 63°53′ W. longitude. Air Rescue Service SA-16 (amphibian) was launched and instructed to proceed to this point.
12:35 p.m.	–	All units in area El were proceeding to calculated landing position.
12:41 p.m.	00:00	Spacecraft landed.
12:44 p.m.	00:03	Contingency recovery situation was established at Recovery Control Center. Recovery commander in area H (embarked on USS *Intrepid*) was designated mission coordinator. Positions of vessels in vicinity of landing point requested from Coast Guard and other naval commands.
12:47 p.m.	00:06	Search aircraft reported possible UHF/DF contact with spacecraft at 04:54 GET (ground elapsed time).
12:58 p.m.	00:17	Destroyer USS *Farragut* proceeding to calculated landing position.
12:59 p.m.	00:18	All search aircraft executing search plan. Positive UHF/DF contact with spacecraft.
1:20 p.m.	00:39	Search aircraft reported visual contact with green dye at 19°29′ N. 64°05′ W. Spacecraft employs fluorosene sea-marker.
1:21 p.m.	00:40	Search aircraft reported astronaut in life raft attached to spacecraft.
1:27 p.m.	00:46	Search aircraft reported that astronaut appeared to be comfortable.
1:34 p.m.	00:53	The SC-54 descended to deploy pararescue team and auxiliary flotation collar.

(continued)

© Springer International Publishing Switzerland 2016

C. Burgess, *Aurora 7*, Springer Praxis Books, DOI 10.1007/978-3-319-20439-0

E.S.T. hr:mn	Elapsed time from landing hr:mn	Event
1:40 p.m.	00:59	Pararescue team deployed.
1:40 p.m.	00:59	Two HSS-2 helicopters launched from USS *Intrepid* with Project Mercury doctor and specially equipped swimmers aboard.
1:50 p.m.	01:09	The SA-16 arrived on scene.
1:56 p.m.	01:15	SA-16 descended to evaluate sea condition for possible landing.
2:15 p.m.	01:34	SA-16 reported sea condition satisfactory for landing and takeoff.
2:21 p.m.	01:40	Astronaut appeared normal, and waved to aircraft. Pararescue team in water. Helicopters en route. SA-16 told not to land unless helicopter retrieval could not be made.
2:39 p.m.	01:58	Auxiliary flotation collar attached to spacecraft and inflated.
2:52 p.m.	02:11	Astronaut and pararescue team in water. No direct communication with astronaut. Astronaut appeared to be in good condition.
3:30 p.m.	02:49	Helicopter arrived over spacecraft.
3:40 p.m.	02:59	Astronaut in helicopter. Doctor reported astronaut in good condition.
3:42 p.m.	03:01	Helicopter retrieved pararescue team. Astronaut Carpenter reported, "Feel fine." Destroyer USS *Farragut* 18 miles from spacecraft.
4:05 p.m.	03:24	Helicopters returned to the USS *Intrepid* accompanied by SA-16 and search aircraft.
4:20 p.m.	03:39	USS *Farragut* had spacecraft in sight.
4:52 p.m.	04:11	Astronaut arrived aboard USS *Intrepid*.
6:16 p.m.	05:35	USS *John R. Pierce* had USS *Farragut* in sight.
6:52 p.m.	06:11	USS *John R. Pierce* had spacecraft onboard.
7:15 p.m.	06:34	Initial medical examination and debriefing of astronaut completed onboard USS *Intrepid*. Astronaut departed for Grand Turk Island.

Appendix 7

Scott Carpenter in popular culture

In the 1960s, one of the most popular comic strips in daily newspaper around the world was the Charles M. Schultz creation, *Peanuts,* which featured the youthful travails of Charlie Brown. On 28 June 1962, a whimsical reference was made to Scott Carpenter's dramatic recovery from the Atlantic Ocean in a strip in which Brown's best friend, Linus Van Pelt, recovers his lost security blanket.

The *Peanuts* strip from 28 June 1962. (Charles M. Schultz, United Feature Syndicate)

© Springer International Publishing Switzerland 2016
C. Burgess, *Aurora 7*, Springer Praxis Books, DOI 10.1007/978-3-319-20439-0

Thunderbirds was a British science fiction series filmed for television between 1964 and 1966. Created by Gerry and Sylvia Anderson, this highly popular show featured marionette puppets and scale-model special effects, and was centered around a futuristic organization that called itself International Rescue (IR). According to the storyline, IR was founded by ex-astronaut Jeff Tracy and his five sons, Alan, John, Gordon, Virgil and Scott (named after five of the Mercury astronauts), who were the heroic pilots for various innovative machines. Scott, the eldest son of Jeff Tracy, was named after Scott Carpenter, although Carpenter must have been somewhat bemused that his marionette character was said to have been a former Air Force pilot, rather than a naval aviator, and that his likeness was based on that of actor Sean Connery.

The marionette Scott Tracy of *Thunderbirds* fame. (Photo: Gerry Anderson/ITV Studios)

The Scott Tracy character also appeared in the 2004 live-action film *Thunderbirds*, based on the earlier TV series, in which the part was played by Phillip Winchester, whose other film roles included *The Patriot* (1998) and *Flyboys* (2006), and he currently has a recurring role in the TV series, *Strike Back*. The voice of Scott Tracy in the television series was supplied by Canadian actor Shane Rimmer.

In the 1983 movie, *The Right Stuff*, based on the seminal book by Tom Wolfe, the part of Mercury astronaut Scott Carpenter was played by American actor, Charles Frank. Directed by Philip Kaufman, the movie was nominated for eight Academy Awards, including Best Picture, and in the end picked up four Oscars. Although the film was based upon the exploits of all seven Mercury astronauts, the role played by Charles Frank was a relatively minor one.

Published in 1992, Thomas Mallon's book *Aurora 7* is a novel in which the story unfolds on 24 May 1962, the day Scott Carpenter flew the MA-7 mission. Mallon once

Actor Charles Frank portrayed Scott Carpenter in *The Right Stuff* (Photo: Warner Brothers)

described it as "a little collage of things happening moment-by-moment during a five-hour space flight." The book tells the story of a young boy named Gregory Noonan who is so enthralled by the flight of *Aurora 7* that he skips school and makes his way into New York City, where crowds have gathered in Grand Central Station to watch the launch live from Cape Canaveral on a huge television screen. As the plot unfolds, various characters find themselves in different situations as Scott Carpenter continues to orbit the Earth.

About the author

Australian author Colin Burgess grew up in Sydney's southern suburbs where he and his wife Patricia still live. They have two grown sons, two grandsons and a granddaughter.

His working life began in the wages department of a major Sydney afternoon newspaper, where he first picked up his writing bug, and later as a sales representative for a precious metals company. He subsequently joined Qantas Airways as a passenger handling agent in 1970 and two years later transferred to the airline's cabin crew. He retired from Qantas as an onboard Flight Service Director/Customer Service Manager in 2002, after 32 years' service.

During that time, he published several books on the experiences of Australian prisoners of war, as well as biographies on space explorers such as Australian payload specialist Dr. Paul Scully-Power and Challenger teacher Christa McAuliffe. He has also written extensively on space flight subjects for astronomy and space-related magazines in Australia, the United Kingdom and the Unites States. In 2003 the University of Nebraska Press appointed him Series Editor for their ongoing *Outward Odyssey* series of books detailing the entire social history of space exploration, and he was involved in co-writing three of these volumes.

His first Springer-Praxis book, *NASA's Scientist-Astronauts*, which was co-authored with British-based space historian David J. Shayler, was released in 2007. *Aurora 7* will be his ninth title with Springer-Praxis, for whom he is currently researching further books for future publication. He regularly attends astronaut functions in the United States and is well known to many of the pioneering space explorers, allowing him to conduct personal interviews for these books.

© Springer International Publishing Switzerland 2016

C. Burgess, *Aurora 7*, Springer Praxis Books, DOI 10.1007/978-3-319-20439-0

Index

© Springer International Publishing Switzerland 2016
C. Burgess, *Aurora 7*, Springer Praxis Books, DOI 10.1007/978-3-319-20439-0